Putting **Essential Understanding** of

Geometry
into Practice

W0038246

in Grades
6–8

Terry Crites
Northern Arizona University
Flagstaff, Arizona

Barbara J. Dougherty
University of Hawaii
Honolulu, Hawaii

Hannah Slovin
University of Hawaii
Honolulu, Hawaii

Karen Karp
Johns Hopkins University
Baltimore, Maryland

Volume Editor
Barbara J. Dougherty
Series Editor
University of Hawaii
Honolulu, Hawaii

NATIONAL COUNCIL OF
TEACHERS OF MATHEMATICS

www.nctm.org/more4u
Access code: RAP14349

Copyright © 2018 by
The National Council of Teachers of Mathematics, Inc.
1906 Association Drive, Reston, VA 20191-1502
(703) 620-9840; (800) 235-7566; www.nctm.org
All rights reserved

Library of Congress Cataloging-in-Publication Data

Names: Crites, Terry, author. | Dougherty, Barbara J., author. | Slovin,
 Hannah, author. | Karp, Karen S., author.
Title: Putting essential understanding of geometry into practice in grades 6-8 /
 Terry Crites (Northern Arizona University, Flagstaff, Arizona),
 Barbara Dougherty (University of Hawaii, Honolulu, HI), Hannah Slovin
 (University of Hawaii, Honolulu, Hawaii), Karen Karp (Johns Hopkins University,
 Baltimore, Maryland).
Description: Reston, VA : National Council of Teachers of Mathematics, ©2017.
 | Series: Putting essential understanding into practice series | Includes
bibliographical references.
Identifiers: LCCN 20170 41599 (print) I LCCN 2017053473 (ebook) | ISBN 9780873539197
 (ebook) | ISBN 9780873537346 (pbk.)
Subjects: LCSH: Geometry--Study and teaching (Elementary) I Geometry-Study
 and teaching (Middle school) | Measurement--Study and teaching (Elementary)
 | Measurement--Study and teaching (Middle school)
Classification: LCC QA465 (ebook) I LCCQA465 .C75 2017 (print)
 | DOC 372.7/ 6--dc23
LC record available at https:lccn.loc.gov/2017041599

The National Council of Teachers of Mathematics supports and advocates for the
highest-quality mathematics teaching and learning for each and every student.

Printed in the United States of America

Contents

Chapter 1

Geometric Measurement

Chapter 2

Transformations

Chapter 3
Looking Back and Looking Ahead with Geometry

Appendix 1
The Big Ideas and Essential Understandings for Geometry

Appendix 2
Resources for Teachers

Appendix 3
Tasks

References

Accompanying Materials at More4U

Appendix 1
The Big Ideas and Essential Understandings for Geometry X

Appendix 2
Resources for Teachers

Appendix 3
Tasks

Foreword

Teaching mathematics in prekindergarten–grade 12 requires knowledge of mathematical content and developmentally appropriate pedagogical knowledge to provide students with experiences that help them learn mathematics with understanding, while they reason about and make sense of the ideas that they encounter.

In 2010 the National Council of Teachers of Mathematics (NCTM) published the first book in the Essential Understanding Series, focusing on topics that are critical to the mathematical development of students but often difficult to teach. Written to deepen teachers' understanding of key mathematical ideas and to examine those ideas in multiple ways, the Essential Understanding Series was designed to fill gaps and extend teachers' understanding by providing a detailed survey of the big ideas and the essential understandings related to particular topics in mathematics.

The Putting Essential Understanding into Practice Series builds on the Essential Understanding Series by extending the focus to classroom practice. These books center on the pedagogical knowledge that teachers must have to help students master the big ideas and essential understandings at developmentally appropriate levels.

To help students develop deeper understanding, teachers must have skills that go beyond knowledge of content. The authors demonstrate that for teachers—

- understanding student misconceptions is critical and helps in planning instruction;

- knowing the mathematical content is not enough—understanding student learning and knowing different ways of teaching a topic are indispensable; and

- constructing a task is important because the way in which a task is constructed can aid in mediating or negotiating student misconceptions by providing opportunities to identify those misconceptions and determine how to address them.

Through detailed analysis of samples of student work, emphasis on the need to understand student thinking, suggestions for follow-up tasks with the potential to move students forward, and ideas for assessment, the Putting Essential Understanding into Practice Series demonstrates best practice for developing students' understanding of mathematics.

The ideas and understandings that the Putting Essential Understanding into Practice Series highlights for student mastery are also embodied in the Common Core State Standards for Mathematics, and connections with these new standards are noted throughout each book.

On behalf of the Board of Directors of NCTM, I offer sincere thanks to everyone who has helped to make this new series possible. Special thanks go to Barbara J. Dougherty for her leadership as series editor and to all the authors for their work on the Putting Essential Understanding into Practice Series. I join the project team in welcoming you to this special series and extending best wishes for your ongoing enjoyment—and for the continuing benefits for you and your students—while you explore Putting Essential Understanding into Practice!

Linda M. Gojak
President, 2012–2014
National Council of Teachers of Mathematics

Preface

The Putting Essential Understanding into Practice Series explores the teaching of mathematics topics in K–grade 12 that are difficult to learn and to teach. Each volume in this series focuses on specific content from one volume in NCTM's Essential Understanding Series and links it to ways in which those ideas can be taught successfully in the classroom.

Thus, this series builds on the previous series, which aimed to present the mathematics that teachers need to know and understand well to teach challenging topics successfully to their students. Each of the earlier books identified and examined the big ideas related to the topic, as well as the "essential understandings"–the associated smaller, and often more concrete, concepts that compose each big idea.

Taking the next step, the Putting Essential Understanding into Practice Series shifts the focus to the specialized pedagogical knowledge that teachers need to teach those big ideas and essential understandings effectively in their classrooms. The Introduction to each volume details the nature of the complex, substantive knowledge that is the focus of these books–*pedagogical content knowledge.* For the topics explored in these books, this knowledge is student centered and focuses on teaching mathematics through problem solving.

Each book then puts big ideas and essential understandings related to the topic under a high-powered teaching lens, showing in fine detail how they might be presented, developed, and assessed in the classroom. Specific tasks, classroom vignettes, and samples of student work illustrate possible methods of introducing students to the ideas in ways that will enable students not only to make sense of them now but also to build on them in the future. Items for readers' reflection appear throughout and offer teachers additional opportunities for professional development.

The final chapter of each book looks at previous and subsequent instruction on the topic. A look back highlights effective teaching that lays the earlier foundations that students are expected to bring to the current grades, where they solidify and build on previous learning. A look ahead reveals how high-quality teaching can expand students' understanding when they move to more advanced levels.

Each volume in the Putting Essential Understanding into Practice Series also includes three appendixes to extend and enrich readers' experiences and possibilities for using the book. The appendixes list the big ideas and essential understandings related to the topic, detail resources for teachers, and present a selection of tasks discussed in the book. These materials are also available to readers online at the More4U website, where Appendix 3 includes additional tasks, and all the tasks are print-ready to facilitate hands-on work with students. Readers can gain online access to each book's More4U materials by going to www.nctm.org/more4u and entering the code that appears on the title page. They can then print out these materials for personal or classroom use.

Because the topics chosen for both the Essential Understanding Series and this successor series represent areas of mathematics that are widely regarded as challenging to teach and to learn, we believe that these books fill a tangible need for teachers. We hope that as you move through the tasks and consider the associated classroom implementations, you will find a variety of ideas to support your teaching and your students' learning.

Acknowledgments from the Authors

We wish to express our thanks for the various contributions to the development of the manuscript of this book from the following: Students at University Laboratory School, Honolulu, Hawaii; Brendan Brennan, University Laboratory School, Honolulu, Hawaii; Stephanie Capen, Honolulu, Hawaii; and Maryam Abhari, Honolulu, Hawaii.

Introduction

Shulman (1986, 1987) identified seven knowledge bases that influence teaching:

1. Content knowledge

2. General pedagogical knowledge

3. Curriculum knowledge

4. Knowledge of learners and their characteristics

5. Knowledge of educational contexts

6. Knowledge of educational ends, purposes, and values

7. Pedagogical content knowledge

The specialized content knowledge that you use to transform your understanding of mathematics content into ways of teaching is what Shulman identified as item 7 on this list—*pedagogical content knowledge* (Shulman 1986). This is the knowledge that is the focus of this book—and all the volumes in the Putting Essential Understanding into Practice Series.

Pedagogical Content Knowledge

In mathematics teaching, pedagogical content knowledge includes at least four indispensable components:

1. Knowledge of curriculum for mathematics

2. Knowledge of assessments for mathematics

3. Knowledge of instructional strategies for mathematics

4. Knowledge of student understanding of mathematics (Magnusson, Krajcik, and Borko 1999)

These four components are linked in significant ways to the content that you teach.

Even though it is important for you to consider how to structure lessons, deciding what group and class management techniques you will use, how you will allocate time, and what will be the general flow of the lesson, Shulman (1986) noted that it is even more important to consider *what* is taught and the *way* in which it is taught.

Every day, you make at least five essential decisions as you determine—

1. which explanations to offer (or not);

2. which representations of the mathematics to use;

3. what types of questions to ask;

4. what depth to expect in responses from students to the questions posed; and

5. how to deal with students' misunderstandings when these become evident in their responses.

Your pedagogical content knowledge is the unique blending of your content expertise and your skill in pedagogy to create a knowledge base that allows you to make robust instructional decisions. Shulman (1986, p. 9) defined pedagogical content knowledge as "a second kind of content knowledge. . . , which goes beyond knowledge of the subject matter per se to the dimension of subject matter knowledge *for teaching.*" He explained further:

> Pedagogical content knowledge also includes an understanding of what makes the learning of specific topics easy or difficult: the conceptions and preconceptions that students of different ages and backgrounds bring with them to the learning of those most frequently taught topics and lessons. (p. 9)

If you consider the five decision areas identified at the top of the page, you will note that each of these requires knowledge of the mathematical content and the associated pedagogy. For example, you may encounter students who are confused about the relationship between perimeter and area and believe that knowing the perimeter allows them to predict the area, and vice versa. Your knowledge of geometry and measurement can help you craft tasks and questions that provide counterexamples and ways to guide your students in seeing connections within and among the multiple aspects of geometry and measurement concepts. As you establish the content, complete with learning goals, you need to consider how to move your students from their initial understandings to deeper ones, building rich connections along the way.

The instructional sequence that you design to meet student learning goals has to take into consideration the misconceptions and misunderstandings that you might expect to encounter (along with the strategies that you expect to use to negotiate them), your expectation of the level of difficulty of the topic for your students, the progression of experiences in which your students will engage, appropriate collections of representations for the content, and relationships between and among geometry and other topics.

Model of Teacher Knowledge

Grossman (1990) extended Shulman's ideas to create a model of teacher knowledge with four domains (see fig. 0.1):

1. Subject-matter knowledge

2. General pedagogical knowledge

3. Pedagogical content knowledge

4. Knowledge of context

Subject-matter knowledge includes mathematical facts, concepts, rules, and relationships among concepts. Your understanding of the mathematics affects the way in which you teach the content—the ideas that you emphasize, the ones that you do not, the algorithms that you use, and so on (Hill, Rowan, and Ball 2005).

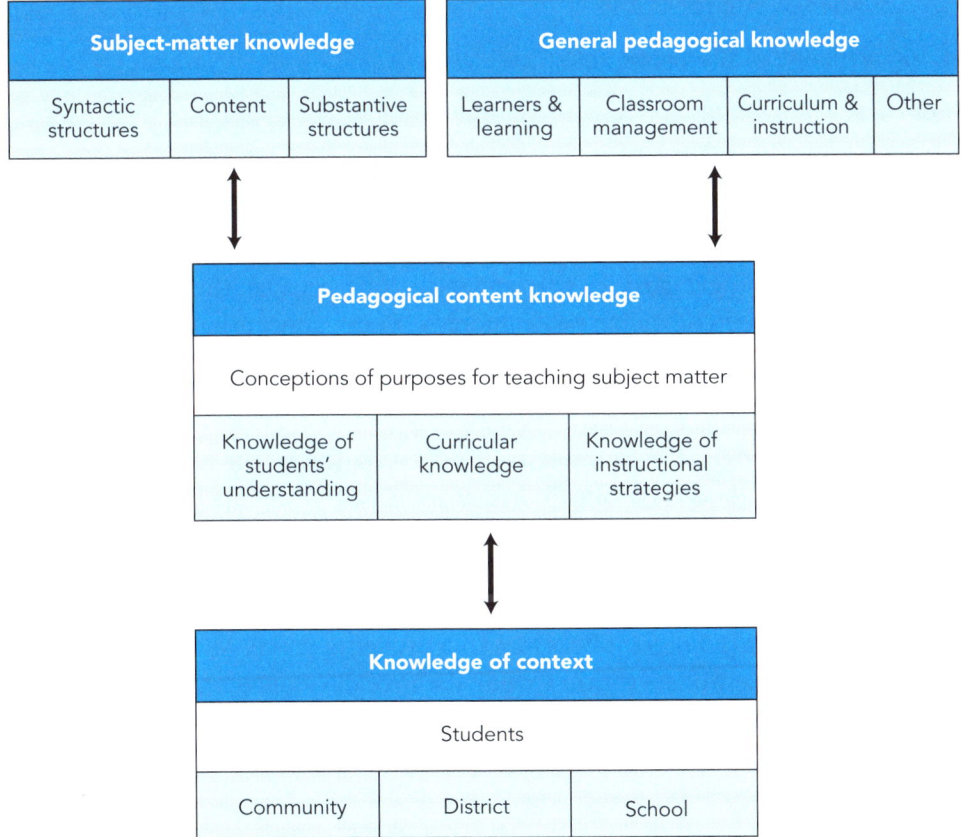

Fig. 0.1. Grossman's (1990, p. 5) model of teacher knowledge

Your pedagogical knowledge relates to the general knowledge, beliefs, and skills that you possess about instructional practices. These include specific instructional strategies that you use, the amount of wait time that you allow for students' responses to questions or tasks, classroom management techniques that you use for setting expectations and organizing students, and your grouping techniques, which might include having your students work individually or cooperatively, collaboratively, in groups or pairs. As Grossman's model indicates, your understanding and interpretation of the environment of your school, district, and community can also affect the way in which you teach a topic.

Note that pedagogical content knowledge has four aspects, or components, in Grossman's (1990) model:

1. Conceptions of purposes for teaching subject matter

2. Knowledge of students' understanding

3. Knowledge of curriculum

4. Knowledge of instructional strategies

Each of these components has specific connections to the classroom. It is useful to consider each one in turn.

First, when you think about the goals that you want to establish for your instruction, you are focusing on your conceptions of the purposes for teaching subject matter. This is a broad category but an important one because the goals that you set will define learning outcomes for your students. These conceptions influence the other three components of pedagogical content knowledge. Hence, they appropriately occupy their overarching position in the model.

Second, your knowledge of your students' understanding of the mathematics content is central to good teaching. To know what your students understand, you must focus on both their conceptions and their misconceptions. As teachers, we all recognize that students develop naïve understandings that may or may not be immediately evident to us in their work or discourse. These can become deep-rooted misconceptions that are not simply errors that students make. Misconceptions may include incorrect generalizations that students have developed, such as the idea that area is the product of the lengths of the base and the height, without the complementary insight that this is true only for parallelograms. These generalizations may even be predictable notions that students exhibit as part of a developmental trajectory, as they move from counting squares in rectangular arrays to finding the area of rectangles.

Part of your responsibility as a teacher is to present tasks or to ask questions that can bring misconceptions to the forefront. Once you become aware of misconceptions in students' thinking, you then have to determine the next instructional steps. The mathematical ideas presented in this volume focus on common misconceptions that students form in relation to a specific topic—geometry in grades 6–8. This book shows how the type of task selected and the sequencing of carefully developed questions can bring the misconceptions to light, as well as how particular teachers took the next instructional steps to challenge the students' misconceptions.

Third, curricular knowledge for mathematics includes multiple areas. Your teaching may be guided by a set of standards such as the Common Core State Standards for Mathematics (CCSSM; National Governors' Association Center for Best Practices and Council of Chief State School Officers 2010) or other provincial, state, or local standards. You may in fact use these standards as the expected learning outcomes for your students. Your textbook is another source that may influence your instruction. With any textbook also comes a particular philosophical view of mathematics, mathematics teaching, and student learning. Your awareness and understanding of the curricular perspectives related to the choice of standards and the selection of a textbook can help determine how you actually enact your curriculum. Moreover, your district or school may have a pacing guide that influences your delivery of the curriculum. In this book, we can focus only on the alignment of the topics presented with broader curricular perspectives, such as CCSSM. However, your own understanding of and expertise with your other curricular resources, coupled with the parameters defined by the expected student outcomes from standards documents, can provide the specificity that you need for your classroom.

In addition to your day-to-day instructional decisions, you make daily decisions about which tasks from curricular materials you can use without adaptation, which tasks you need to adapt, and which tasks you need to create on your own. Once you select or develop meaningful, high-quality tasks and use them in your mathematics lesson, you have launched what Yinger (1988) called "a three-way conversation between teacher, student, and problem" (p. 86). This process is not simple—it is complex because how students respond to the problem or task is directly linked to your next instructional move. That means that you have to plan multiple instructional paths to choose among as students respond to those tasks.

Knowledge of the curriculum goes beyond the curricular materials that you use. You also consider the mathematical knowledge that students bring with them from grade 5 and what they should learn by the end of grade 8. The way in which you teach a foundational concept or skill has an impact on the way in which students will interact with and learn subsequent related content. For example, the types of representations that you include in your introduction of geometry are the ones that your students will use to evaluate other representations and ideas in later grades.

Fourth, knowledge of instructional strategies is essential to pedagogical content knowledge. Having a wide array of instructional strategies for teaching mathematics is central to effective teaching and learning. Instructional strategies, along with knowledge of the curriculum, may include the selection of mathematical tasks, together with the way in which those tasks are enacted in the classroom. Instructional strategies may also include the way in which the mathematical content is structured for students. You may have very specific ways of thinking about how you will structure your presentation of a mathematical idea—not only how you will sequence the introduction and development of the idea, but also how you will present that idea to your students. Which examples should you select, and which questions should you ask? What representations should you use? Your knowledge of instructional strategies, coupled with your knowledge of your curriculum, permits you to align the selected mathematical tasks closely with the way in which your students perform those tasks in your classroom.

The instructional approach in this volume combines a student-centered perspective with an approach to mathematics through problem solving. A student-centered approach is characterized by a shared focus on student and teacher conversations, including interactions among students. Students who learn through such an approach are active in the learning process and develop ways of evaluating their own work and one another's in concert with the teacher's evaluation.

Teaching through problem solving makes tasks or problems the core of mathematics teaching and learning. The introduction to a new topic consists of a task that students work through, drawing on their previous knowledge while connecting it with new ideas. After students have explored the introductory task (or tasks), their consideration of solution methods, the uniqueness or multiplicity of solutions, and extensions of the task create rich opportunities for discussion and the development of specific mathematical concepts and skills.

By combining the two approaches, teachers create a dynamic, interactive, and engaging classroom environment for their students. This type of environment promotes the ability of students to demonstrate CCSSM's Standards for Mathematical Practice while learning the mathematics at a deep level.

The chapters that follow show that instructional sequences embed all the characteristics of knowledge of instructional strategies that Grossman (1990) identifies. One component that is not explicit in Grossman's model but is included in a model developed by Magnusson, Krajcik, and Borko (1999) is the knowledge of assessment. Your knowledge of assessment in mathematics plays an important role in guiding your instructional decision-making process.

There are different types of assessments, each of which can influence the evidence that you collect as well as your view of what students know (or do not know) and how they know what they do. Your interpretation of what students know is also related to your view of what constitutes "knowing" in mathematics. While you examine the tasks, classroom vignettes, and samples of student work in this volume, you will notice that teacher questioning permits formative assessment that supplies information that spans both conceptual and procedural aspects of understanding. *Formative assessment*, as this book uses the term, refers to an appraisal that occurs during an instructional segment, with the aim of adjusting instruction to meet the needs of students more effectively (Popham 2006). Formative assessment does not always require a paper-and-pencil product but may include questions that you ask or tasks that students complete during class.

The information that you gain from student responses can provide you with feedback that guides the instructional flow, while giving you a sense of how deeply (or superficially) your students understand a particular idea—or whether they hold a misconception that is blocking their progress. As you monitor your students' development of rich understanding, you can continually compare their responses with your expectations and then adapt your instructional plans to accommodate their current levels of development. Wiliam (2007, p. 1054) described this interaction between teacher expectations and student performance in the following way:

> It is therefore about assessment functioning as a bridge between teaching and learning, helping teachers collect evidence about student achievement in order to adjust instruction to better meet student learning needs, in real time.

Wiliam notes that for teachers to get the best information about student understandings, they have to know how to facilitate substantive class discussions, choose tasks that include opportunities for students to demonstrate their learning, and employ robust and effective questioning strategies. From these strategies, teachers must then interpret student responses and scaffold their learning to help them progress to more complex ideas.

Characteristics of Tasks

The type of task that is presented to students is very important. Tasks that focus only on procedural aspects may not help students learn a mathematical idea deeply. Superficial learning may result in students forgetting easily, requiring reteaching, and potentially affecting how they understand mathematical ideas that they encounter in the future. Thus, the tasks selected for inclusion in this volume emphasize deep learning of significant mathematical ideas. These rich, "high-quality" tasks have the power to create a foundation for more sophisticated ideas and support an understanding that goes beyond "how" to "why." Figure 0.2 identifies the characteristics of a high-quality task.

While you move through this volume, you will notice that it sequences tasks for each mathematical idea so that they provide a cohesive and connected approach to the identified concept. The tasks build on one another to ensure that each student's thinking becomes increasingly sophisticated, progressing from a novice's view of the content to a perspective that is closer to that of an expert. We hope that you will find the tasks useful in your own classes.

A high-quality task has the following characteristics:
Aligns with relevant mathematics content standard(s)
Encourages the use of multiple representations
Provides opportunities for students to develop and demonstrate the mathematical practices
Involves students in an inquiry-oriented or exploratory approach
Allows entry to the mathematics at a low level (all students can begin the task) but also has a high ceiling (some students can extend the activity to higher-level activities)
Connects previous knowledge to new learning
Allows for multiple solution approaches and strategies
Engages students in explaining the meaning of the result
Includes a relevant and interesting context

Fig. 0.2. Characteristics of a high-quality task

Types of Questions

The questions that you pose to your students in conjunction with a high-quality task may at times cause them to confront ideas that are at variance with or directly contradictory to their own beliefs. The state of mind that students then find themselves in is called *cognitive dissonance,* which is not a comfortable state for students—or, on occasion, for the teacher. The tasks in this book are structured in a way that forces students to deal with two conflicting ideas. However, through the process of negotiating the contradictions, students come to know the content much more deeply. How the teacher handles this negotiation determines student learning.

You can pose three types of questions to support your students' process of working with and sorting out conflicting ideas. These questions are characterized by their potential to encourage reversibility, flexibility, and generalization in students' thinking (Dougherty 2001). All three types of questions require more than a one-word or one-number answer. Reversibility questions are those that have the capacity to change the direction of students' thinking. They often give students the solution and require them to create the corresponding problem. A flexibility question can be one of two types: it can ask students to solve a problem in more than one way, or it can ask them to compare two or more problems or determine the relationship between or among concepts and skills. Generalization questions also come in two types: they ask students to look at multiple examples or cases and find a pattern or make observations, or they ask them to create a specific example of a rule, conjecture, or pattern. Figure 0.3 provides examples of reversibility, flexibility, and generalization questions related to geometry in grades 6–8.

Type of question	Example
Reversibility question	What are the dimensions of three rectangles, each with a perimeter of 48 units?
Flexibility question	What is the area of the trapezoid? Find the area in two different ways.
Flexibility question	How is finding the area of a rectangle similar to or different from finding the area of a parallelogram?
Generalization question	What remains the same in all rotations? What changes? Explain.
Generalization question	Jackson said, "As the surface area of a rectangular prism increases, the volume of the prism also increases." Do you agree with Jackson? Support your answer with examples.

Fig. 0.3. Examples of reversibility, flexibility, and generalization questions

Conclusion

The Introduction has provided a brief overview of the nature of—and necessity for—pedagogical content knowledge. This knowledge, which you use in your classroom every day, is the indispensable medium through which you transmit your understanding of the big ideas of the mathematics to your students. It determines your selection of appropriate, high-quality tasks and enables you to ask the types of questions that will not only move your students forward in their understanding but will also allow you to determine the depth of that understanding.

The chapters that follow describe important ideas related to learners, curricular goals, instructional strategies, and assessment that can assist you in transforming your students' knowledge into formal mathematical ideas related to geometry. These chapters provide specific examples of mathematical tasks and student thinking for you to analyze to develop your pedagogical content knowledge for teaching geometry in grades 6–8 or to give you ideas to help other colleagues develop this knowledge. You will also see how to bring together and interweave your knowledge of learners, curriculum, instructional strategies, and assessment to support your students in grasping the big ideas and essential understandings and using them to build more sophisticated knowledge.

Students in grades 6–12 have already had some experiences that affect their initial understanding of geometry in these grades. Furthermore, they have developed some ideas about geometry in a variety of contexts. Students in the upper elementary grades frequently demonstrate understanding of mathematical ideas related to geometry in a particular context or in connection with related topics. Yet, in other situations, these same students do not demonstrate that same understanding. As a teacher in the middle grades, you must understand the ideas that your students have developed about geometry in their previous experiences so that you can extend their knowledge and see whether or how it differs from the formal mathematical knowledge that they need to be successful in reasoning with or applying geometry. You have the important responsibility of assessing their current knowledge related to the big ideas of geometry as well as their understanding of various representations of these ideas and their power and limitations. Your understanding will facilitate and reinforce your instructional decisions. Teaching the big mathematical ideas and helping students develop essential understandings related to geometry are obviously very challenging and complex tasks.

practice

Chapter 1
Geometric Measurement

Big Idea 1
Behind every measurement formula lies a geometric result.

Essential Understanding 1a
Decomposing and rearranging provide a geometric way of both seeing that a measurement formula is the right one and seeing why it is the right one.

Big Idea 4
Classifying, naming, defining, posing, conjecturing, and justifying are codependent activities in geometric investigation.

The big ideas that teachers need to understand for teaching geometry in the middle grades are explored in *Developing Essential Understanding of Geometry for Teaching Mathematics in Grades 6 8* (Sinclair, Pimm, and Skelin 2012). With each big idea, the authors delineate two or three associated, subordinate ideas—the essential understandings that support the broader concept. Altogether, Sinclair, Pimm, and Skelin identify four big ideas and ten essential understandings that flesh them out (Appendix 1 provides a complete list). Clearly, middle-grades teachers need to have more than a superficial grasp of these big ideas and their related essential understandings before approaching the teaching of geometry in the classroom. A thorough, robust understanding of each one is necessary to prepare teachers to offer rich, effective instruction to students in grades 6–8.

Because teachers' own understanding must extend beyond the concepts and skills that teachers expect their students to learn, key questions arise for classroom instruction:

- Which of the big ideas and essential understandings outlined by Sinclair, Pimm, and Skelin for teachers are critical for middle-grades students to develop while they explore geometry in a more focused and sophisticated way than they did in the elementary grades?

- At what level should students understand these ideas in the middle grades to be ready for the more advanced work with geometry that they will undertake in high school and beyond?

- How can teachers nurture these ideas and understandings most appropriately and effectively in the middle-grades classroom?

These are the questions that this book sets out to answer.

Research and experience provide invaluable assistance in this endeavor, pointing to a number of pedagogical practices that are particularly effective for developing students' mathematical understanding. Useful practices include the following:

- Challenging misconceptions that students commonly form

- Having students analyze one another's work

- Inviting students to describe and critique the thinking of other, sometimes fictitious, students

- Thoughtfully creating cognitive dissonance in students' thinking

- Purposefully selecting counterexamples for students to consider

These are among the practices that this book emphasizes in sample tasks that illustrate ways in which you can develop your middle-grades students' understanding of the core ideas of geometry.

Although this book is not intended to be a comprehensive treatment of middle-grades geometry, it presents tasks that illustrate how you can help your students develop their own conceptual understanding. Each task is designed to provide a rich learning opportunity for students while offering you a window on their thinking. The insights into their misconceptions that these tasks will give you are intended to show you the ideas—large and small—that they do understand, as well as those that they do not understand, thereby supplying you with the information that you need to shape subsequent instruction and tailor it to your students' needs.

While your students complete and discuss these tasks, you will be able to ensure that they receive the kind of feedback that Hattie (2009) claims is one of the most important for the development of student learning. He is not describing teacher feedback regarding student performance but rather students' own feedback to one another on what they know and can apply. This feedback not only provides observant teachers with valuable information about students' understanding of concepts and skills but also prompts students to think about what they have done, thus helping them to self-assess their work.

Not surprisingly, designing such tasks and preparing to present them involves one of the key pedagogical practices for supporting the productive use of students' thinking. This is the instructional practice of anticipating students' solutions—successful and unsuccessful—to set up opportunities to assess and advance that thinking (Stein et al. 2008; Smith and Stein 2011).

Chapter 1 explores topics in geometric measurement. It highlights ways to capture student thinking by using critical pedagogical components. These components include, but are not limited to, the following:

- Determining instructional pathways through the selection and sequencing of activities

- Connecting work in ways that illuminate big ideas of geometry

- Generating a series of questions that have the potential to prompt and probe students' thinking

The first topics that we examine are the concepts of area and perimeter, and the core ideas that come into play in this work are Big Idea 1, Essential Understanding 1*a*, and Big Idea 4.

Working toward Big Idea 1, Essential Understanding 1*a*, and Big Idea 4

Geometric measurement typically lies at the heart of middle-grades investigations of geometry as students expand their understanding of geometric shapes and their attributes and measurement. When students arrive in grades 6–8, they have reached a point in their mathematical development where they can use their growing understanding, together with their emerging skill with algebra, to make generalizations that lead to measurement formulas. Big Idea 1, the notion that behind every measurement formula lies a geometric result, is the central idea that supports and guides this work.

In grades 6–8, students no longer routinely make measurements by counting or working directly with numbers as they did in the elementary grades. To understand geometric measurement more abstractly—to make the connection between a measurement formula and a geometric result—middle-grades students need a powerful tool of geometric comparison: decomposition and composition. As Sinclair, Pimm, and Skelin (2012) express it, "One very powerful means through which such a comparison [without numbers] might occur involves decomposition and re-arrangement" (p. 9). Thus, Essential Understanding 1a is a critical component of Big Idea 1 that middle-grades students must grasp to investigate geometric measurement: "Decomposing and rearranging provide a geometric way of both seeing that a measurement formula is the right one and seeing why it is the right one" (p. 9).

Furthermore, Big Idea 4 naturally underpins students' work with geometric measurement in the middle grades. Big Idea 4 is the notion that classifying, naming, defining, conjecturing, and justifying all play roles in any geometric investigation and are inextricably interrelated. As Sinclair, Pimm, and Skelin (2012) explain, "Geometry arises as the endpoint of geometric investigation" (p. 55), and the actions in Big Idea 4 are "central along the way" because "geometric investigation has a continuity that invokes each of these ... intellectual actions at some point" (p. 55). Chapter 1 focuses particularly on the importance of justifying conjectures to support middle-grades investigations of geometric measurement.

Geometric measurement is a significant area of study for the middle grades, a fact that is underscored by its place in the geometry domain for grades 6–8 in the Common Core State Standards for Mathematics (CCSSM; National Governors Association Center for Best Practices and Council of Chief State School Officers [NGA Center and CCSSO] 2010). Throughout grades 6–8, students are expected to solve real-world and mathematical problems involving the measurement of core attributes of shapes and solids:

> Grade 6 (Geometry, 6.G)
> **Solve real-world and mathematical problems involving area, surface area, and volume.**
>
> 1. Find the area of right triangles, other triangles, special quadrilaterals, and polygons by composing into rectangles or decomposing into triangles and other shapes; apply these techniques in the context of solving real-world and mathematical problems. (p. 44)
>
> Grade 7 (Geometry, 7.G)
> **Solve real-life and mathematical problems involving angle measure, area, surface area, and volume.**
>
> 6. Solve real-world and mathematical problems involving area, volume, and surface area of two- and three-dimensional objects composed of triangles, quadrilaterals, polygons, cubes, and right prisms. (p. 49)

Grade 8 (Geometry, 8.G)

Solve real-world and mathematical problems involving volume of cylinders, cones, and spheres.

> 9. Know the formulas for the volumes of cones, cylinders, and spheres and use them to solve real-world and mathematical problems. (p. 56)

The work of quantifying the attributes of objects in everyday life to make sense of the physical world offers many opportunities for students to meet the expectations of these standards in the CCSSM geometry domain while recognizing their relevance to career options. However, this rich opportunity for application is often ignored. Instead, the typical instructional approach has often been formula-centered, with emphasis on definitions, rules, variables, and dimensions, without adequate attention to students' understanding of these ideas. As Young (1911, pp. 4–5) observed a little more than a century ago,

> The mere memorizing of a demonstration in geometry has about the same education value as the memorizing of a page from the city directory. And yet it must be admitted that a very large number of our pupils do study mathematics in just this way.

Sadly, not much has changed in slightly more than a hundred years, as measured by student performance on major assessments such as the Trends in International Mathematics and Science Study (TIMSS) and National Assessment for Educational Progress (NAEP). Geometric measurement is regularly found to be an area of weakness (Sowder et al. 2004).

Steele (2006) reports that students' misconceptions often lie at the intersection of geometry and measurement. A geometric measurement topic that is especially problematic at the middle-grades level involves the relationships among the measurable quantities of geometric figures, such as area and perimeter.

Interpreting units of length and units of area

The unit of measure is a core concept that underlies both perimeter and area. This is so, in part, because either of these measurements can be found by iterating a unit. For example, the perimeter of a triangle can be determined by identifying a length unit and iterating it around the boundary of the shape. Similarly, the area of a triangle can be found by defining a square unit and iterating it in the interior of the shape. Although the iteration process may not be exact, it can provide at least a good estimate of the measurement in either case. In a broader sense, all measurement requires the integration of spatial and numerical concepts into the unifying idea of an iterated unit, thereby making an emphasis on unitizing critical (Hiebert 1981).

What is more significant for the teaching of area, which usually comes after the teaching of perimeter, is that students tend to confound units of measure for area with units of measure for perimeter. Perimeter, as the measure of the boundary of a shape, is a length, with only one dimension. Unlike perimeter, area, as the measure of the region inside the shape, has two dimensions. Measuring area involves coordinating the measures of the two dimensions of the shape (Outhred and Mitchelmore 2000). In finding the area of a parallelogram, for example, students must coordinate and differentiate two different types of units—linear units when they are finding the length of either of the parallelogram's dimensions or its perimeter, and square units when they are reporting the area enclosed by the parallelogram. In light of students' frequent confusion of linear and square units in the measurement, a critical question for instruction emerges: What would happen if the nature of the unit itself became the focus of consideration in instruction and teachers highlighted the contrast between linear measure and area measure? Linear measurement is an iteration of a linear unit, and area measure is an iteration of an area unit. Reflect 1.1 invites your thinking about this approach in relation to multiplication.

Reflect 1.1

Consider the area of the shape below:

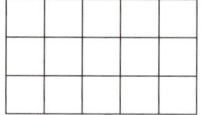

How would thinking of this area of 15 squares as an iteration of a square unit 5 times to make one row unit and then an iteration of the row unit 3 times to make three row units be different from thinking of the same area as length times base?

Are both approaches multiplicative in nature? Why or why not?

An iteration of units suggests an additive approach to measurement because the joining together of the units is indicative of addition. Placing length units on the boundary of a shape fits naturally with perimeter and is closely tied to the additive nature of perimeter formulas. However, covering a shape with square units and counting the number of units to find area does not intuitively align with the multiplicative structure of area formulas.

Multiple researchers—for instance, Battista (2006); Kamii (2006); and Mulligan, Mitchelmore, and Prescott (2005)—have studied students' understanding of unit in relation to perimeter and area. Noticeably fewer researchers have explored the coordination of units across multiple measurement types, such as perimeter and area.

Barrett and colleagues (2011) found that shifting from an additive perspective on units to a multiplicative one requires a purposeful sequencing of tasks that move from qualitative comparisons to quantitative ones, coupled with specific questions focused on the multiplicative relationships.

Outhred and Mitchelmore (2000) proposed a trajectory that moves students from additive thinking to multiplicative structures to develop a relational, rather than an instrumental, understanding of the area formula (Skemp 1978). These researchers found that, for students to progress along this trajectory, they should be able to do the following in sequence (adapted from Outhred and Mitchelmore [2000, p. 161]):

- Completely cover a rectangle by fixed units, without gaps or overlaps

- Spatially structure the units in the array with the same number of units in each row

- Relate both the number of units in each row and the number of rows to the lengths of the sides of the rectangle

- Identify the multiplicative structure of the units in a rectangular array as associated with the number of units in each row and in each column

The last point in Outhred and Mitchelmore's sequence raises an issue: How can students see that this is a multiplicative situation, beyond recognizing that you multiply the dimensions (height × base) in the formula? One way to consider the area of a 3-unit-by-5-unit rectangle is to start with a 1-unit-by-1-unit square, then iterate the square unit 5 times to make one row unit, and then iterate 3 row units to create an area of 15 square units. The area of the 3-unit-by-5-unit rectangle is 15 times larger than the area of the 1-unit-by-1-unit square. The structure of this thinking moves beyond merely substituting the linear dimensions in a formula and instead focuses on the relationships of the unit of measure and the quantity being measured.

Finding perimeter and area, given dimensions

One of the most important responsibilities that teachers have is to create or locate appropriate tasks to advance their students' understanding. Rich tasks can tap into students' thinking and assist them in addressing—and teachers in identifying—their challenges with concepts and procedures that students sometimes understand only in a very fragile way. Such tasks can show both teachers and students what students understand or do not yet understand about the fundamental geometric concepts of perimeter and area. For example, giving students a figure and labeling it in a purposeful way can induce students to demonstrate any weaknesses in their conceptual understanding. Consider the tasks in figure 1.1 as directed by the questions in Reflect 1.2.

Reflect 1.2

Figure 1.1 shows problems 1 and 2, which seek the area and the perimeter, respectively, of a 5-unit by 12-unit rectangle. The rectangle's dimensions are intentionally labeled differently in each case.

How do you suppose the labeling of the rectangle in these problems might affect a student's approach to the problems?

What misconceptions might the student's solutions indicate?

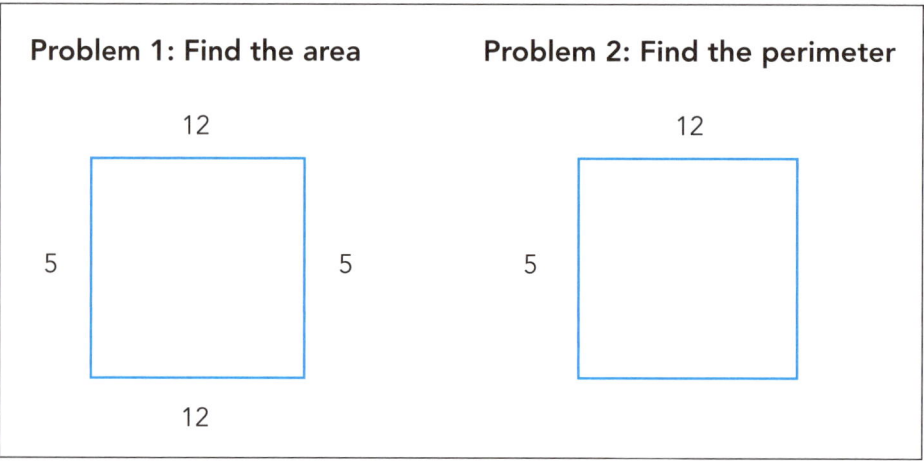

Fig. 1.1. Sample problems with possible labeling of the same rectangle

Samples of student work indicate that the labeling does indeed affect students' solutions. Figure 1.2 shows the work of Anthony, a sixth grader, who attempted to find the area in problem 1 by multiplying all the numbers given by the labels. When asked to explain his work, he said, "I know that when you find area, you have to multiply length and width, so I just multiplied." Clearly, Anthony had difficulty discerning the measures of length and width that he needed. In further discussion, he said that because all the sides were labeled, "That must mean that you have to use all of the measurements."

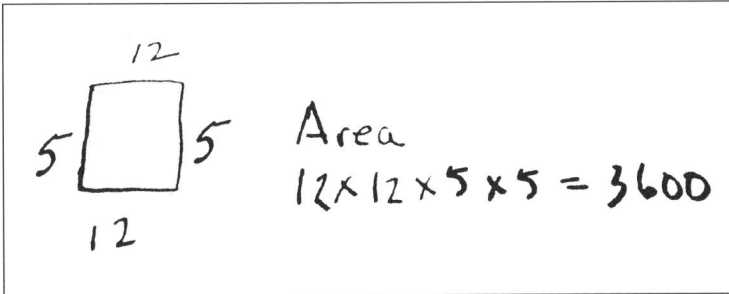

Fig. 1.2. Anthony's (grade 6) solution to problem 1 about the area of the rectangle

Zoie's (a sixth grader) work on problem 2, which seeks the perimeter of the rectangle, is shown in figure 1.3. Using a process similar to that demonstrated by Anthony with respect to the area of the rectangle, Zoie tried to find the rectangle's perimeter by multiplying the two dimensions given by the labels, as shown in the sample of her work in the figure. In oral discussions, she too described the need to use the measurements that were given rather than interpret the meaning of the information in the figure. Neither Zoie nor Anthony considered the reasonableness of their answers. Instead, they focused on selecting and using a formula that seemed to fit the labeling of the shapes.

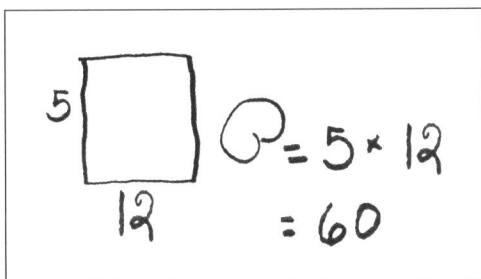

Fig. 1.3. Zoie's (grade 6) solution to problem 2 about the perimeter of the rectangle

The solutions that Zoie and Anthony produced are representative of the various ways in which many students interpret the labeling of figures with respect to a formula. If students are asked to calculate the area of a rectangle that is shown with all four sides labeled, they may opt to use the perimeter formula and find the rectangle's perimeter instead of its area. Or they may seek to apply the area formula but unreflectively. Knowing that this formula involves multiplying a rectangle's dimensions, they may simply multiply all four numbers given. In contrast, if students are asked to find the perimeter of a rectangle that is shown with two adjoining sides labeled, they may resort to using the area formula and find the area instead of the perimeter.

Providing tasks that motivate students to interpret the given information in the context of the problem is important. One of the best ways to do this is to make students stop and think by varying the information given, ranging from giving students the length of one side of a square to giving them rectangles with all four sides labeled in both area and perimeter problems.

When students do not apply a formula or algorithm thoughtfully, deliberately, or prudently, their approach to the formula and their understanding of the relationships expressed in it are more procedural (or instrumental) than relational (Skemp 1978). Stein and colleagues (2008), much like Thompson and colleagues (1994) before them, noted that when the focus of instruction is on the computational aspects of geometric shapes, significant connections between the formulas and the attributes of the actual shapes are missing from students' understanding. By contrast, instruction that varies the information given and focuses the discussion on the relationships between the information given and the formula components can help students develop a stronger conceptual understanding that extends beyond calculations.

Determining the relationship between perimeter and area

While students explore the concepts of perimeter and area, they begin to generalize the relationships between the two. To assist them in this work, they may make comparisons of the perimeters of two shapes if they are given only the areas of the shapes, or vice versa. Consider how students might respond to the task presented in Reflect 1.3.

Reflect 1.3

How do you think your students would respond to the following task:

Carmen said, "As the perimeter of a rectangle increases, the area also increases."

Do you agree with Carmen? Explain your reasoning. (Adapted from Dougherty [2006, p. 56])

It is not uncommon for students to think that a direct relationship exists between perimeter and area. That is, they often generalize that if the perimeter of a given shape increases, the area will also increase. This misconception can be addressed by having students engage in tasks that explicitly direct their attention to the relationship. Consider the two tasks, A and B, shown in figure 1.4, guided by the questions in Reflect 1.4.

Reflect 1.4

Figure 1.4 presents tasks A and B, which are designed to elicit and address students' common misconception that the perimeter and the area of a shape are directly related to each other.

What concepts or big ideas might emerge from discussions about the solutions that students found?

How could those concepts or big ideas be used to address students' misconceptions about the relationship between area and perimeter?

Task A

Use square tiles to create as many different rectangles as possible, each with an area of 36 square units.

Draw your rectangles on grid paper.

Task B

Use square tiles to create as many different rectangles as possible, each with a perimeter of 24 units. Draw your rectangles on grid paper.

Fig. 1.4. Two tasks that focus on the relationship between perimeter and area

One way of using tasks A and B in the classroom is to ask half your students to work on task A and the other half to work on task B. When students complete the tasks, ask those who worked on task A to find the perimeter of each of the rectangles that they created with an area of 36 square units. At the same time, ask those students who worked on task B to find the area of each of the rectangles that they created with a perimeter of 24 units. When you monitor their work, you can ask students to describe their process of determining the perimeters of their shapes in task A. Their responses indicate the level of their thinking in addition to offering you opportunities to link the physical, material representation with a corresponding symbolic representation and develop the relationship in further class discussion. After your students have found all the possible rectangles for their task and have determined the perimeter (task A) or the area (task B), you can ask the groups to describe what they notice. Figures 1.5–1.9 show sample sixth-grade students' observations, paired with the students' drawings of their rectangles.

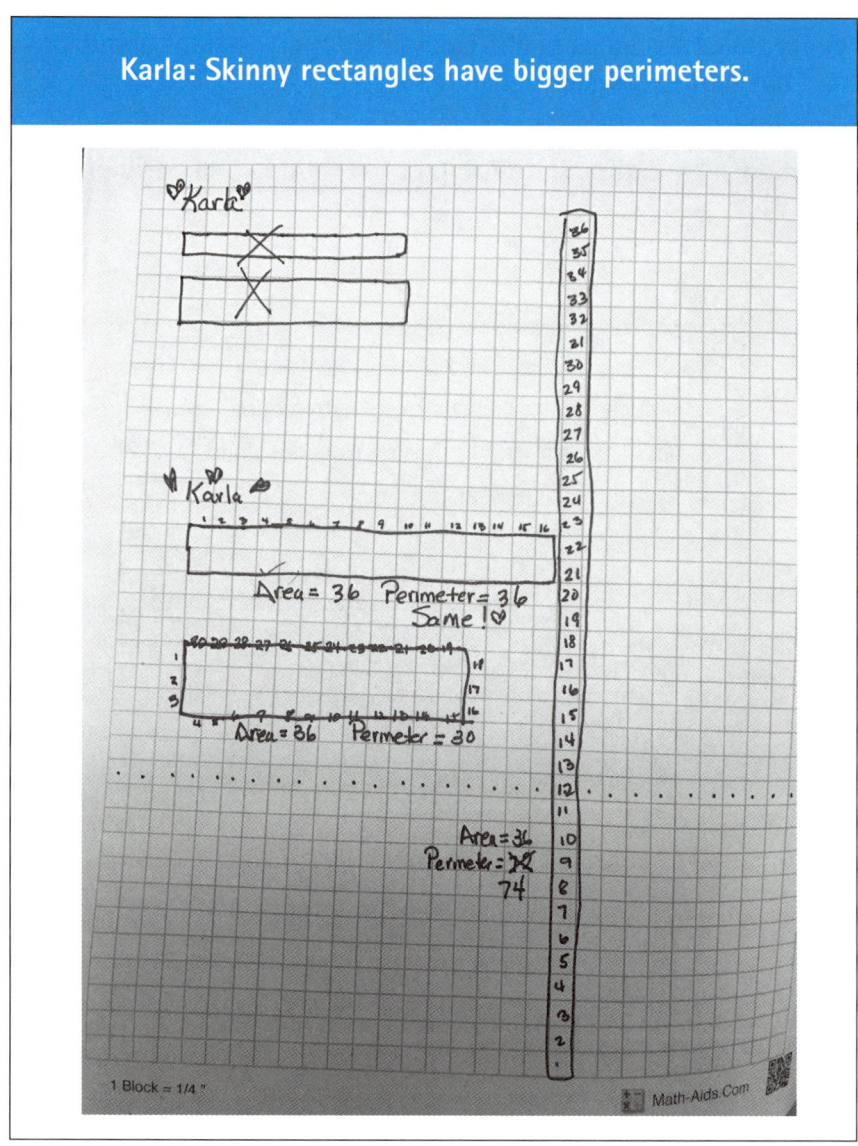

Fig. 1.5. Karla's (grade 6) work and observation on task A

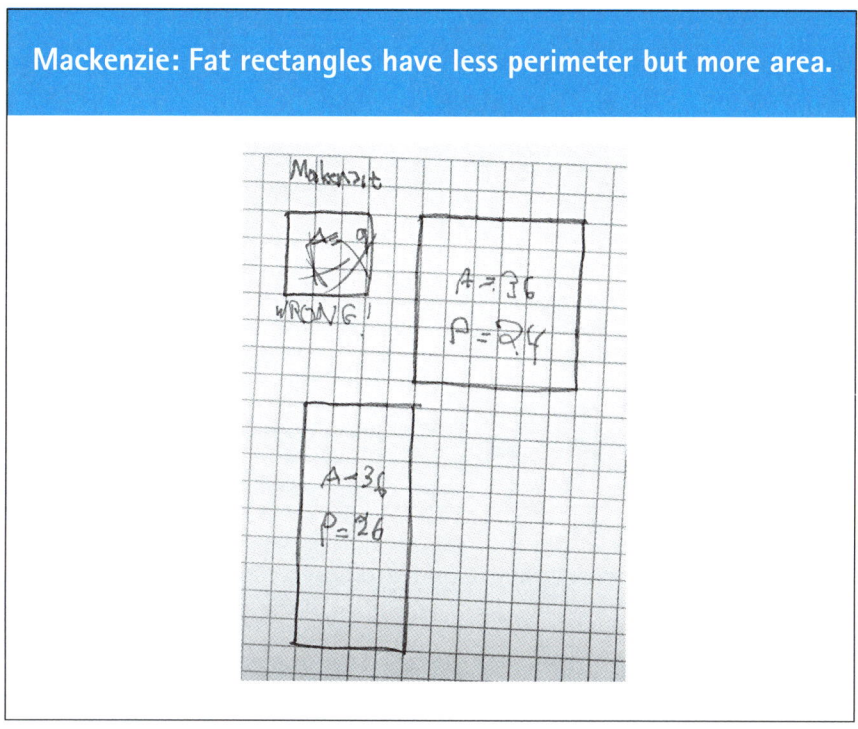

Fig. 1.6. Mackenzie's (grade 6) work and observation on task A

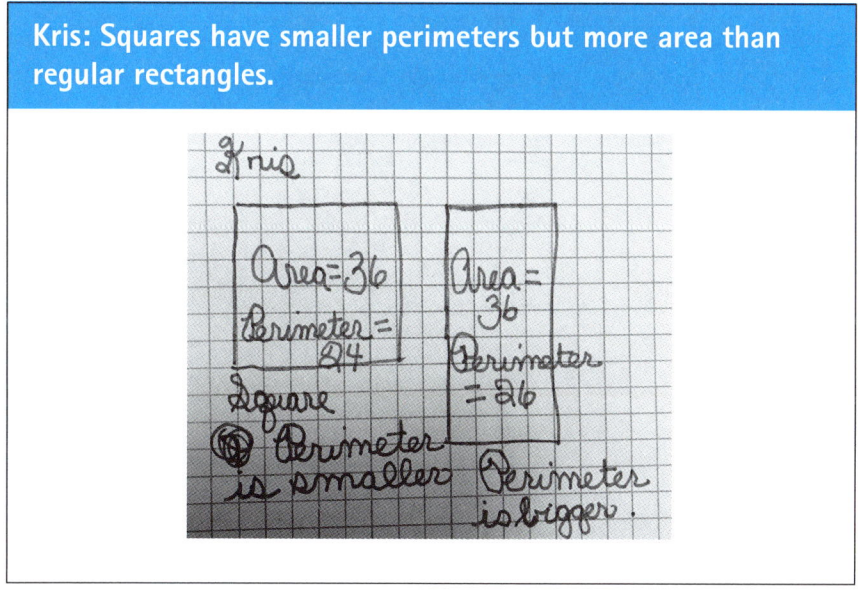

Fig. 1.7. Kris's (grade 6) work and observation on task A

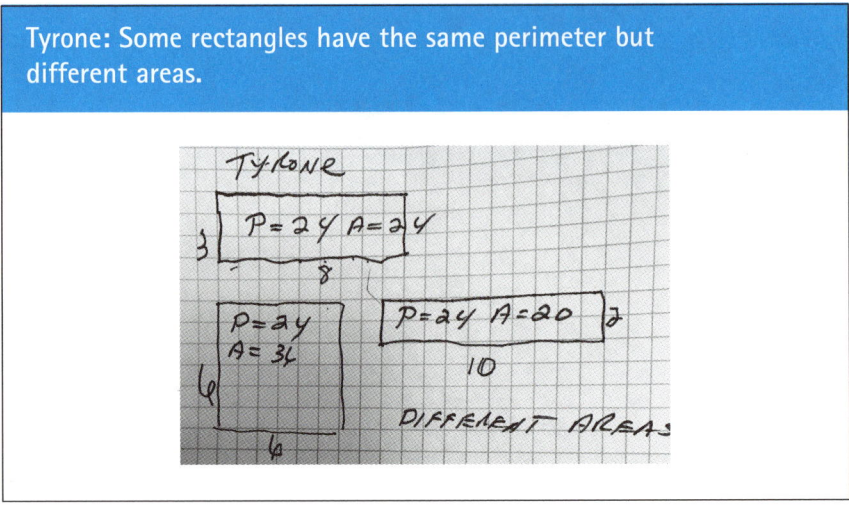

Fig. 1.8. Tyrone's (grade 6) work and observation on task B

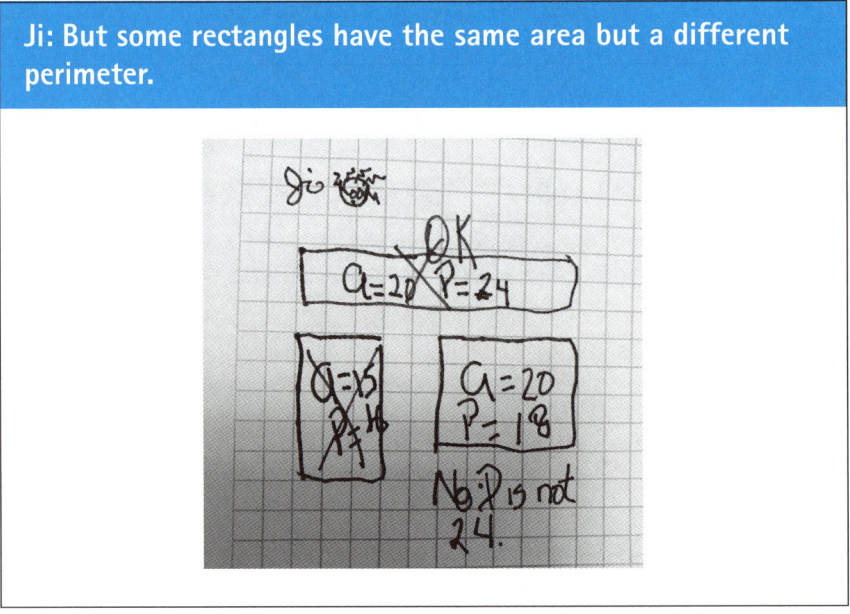

Fig. 1.9. Ji's (grade 6) work and observation on task B

Conjecturing and justifying conjectures are two key activities that Big Idea 4 identifies in geometric investigation, and the class discussion that ensues as students share their observations should include their justifications for the statements that they make. The rationale or support that the students offer for their observations may provide a new perspective that will help other students who have not noticed a particular relationship to see it for the first time or gain further insight into it. The class discussion should also incorporate opportunities for students to ask clarifying questions or furnish further support for the pattern that a student is sharing.

Additionally, the discussion should give students an opportunity to share and examine the processes by which they determined the perimeter and area of the rectangles. To consider the potential variety in these processes, examine the two rectangles in figure 1.10, using the questions in Reflect 1.5 to help you think about the multiple ways in which students could have found the perimeter and area of these rectangles.

Reflect 1.5

Figure 1.10 shows two rectangles on a grid of unit squares. Assuming that students are given rectangles that are presented in this manner—

(a) determine at least three ways in which students might find perimeter;

(b) determine three ways in which students might find area.

How do these methods relate to formulas that can be used to find perimeter or area of rectangles?

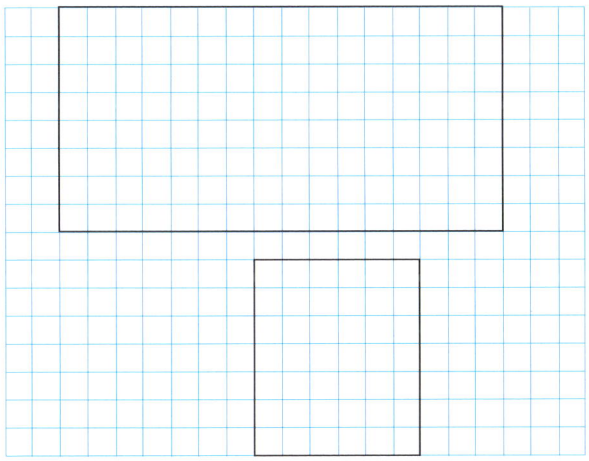

Fig. 1.10. Two rectangles on a grid of unit squares

Students might proceed in multiple ways to find the perimeter of a rectangle like either of those in figure 1.10, composed of unit squares on a square grid. For any of these methods, they would first need to understand that perimeter is a length—the total length of the boundary around a closed figure—and in the case at hand, it is the length of the boundary around the rectangle. These methods include the following:

1. Counting the sides of the unit squares in the grid that adjoin one another to form the boundary of the rectangle.

2. Counting the sides of the unit squares in the grid that adjoin one another to form two adjacent sides of the rectangle and then doubling the count.

3. Counting the sides of the unit squares in the grid that adjoin one another to make up the height of the rectangle, then counting the sides that adjoin one another to make up its base, then adding the two counts, and finally multiplying this sum by 2.

4. Counting the sides of the unit squares in the grid that adjoin one another to form each of the four sides of the rectangle and then adding these four side lengths.

5. Counting the sides of the unit squares in the grid that adjoin one another to make up the height of the rectangle and multiplying this count by 2, then counting the sides that adjoin one another to make up the base of the rectangle and multiplying that count by 2, and finally adding the two products.

Each of these methods can be linked to a symbolic representation, as shown in figure 1.11.

Method	Symbolic representation
Counting the sides of the unit squares in the grid that adjoin one another to form the boundary of the rectangle.	$P = \underbrace{1+1+\cdots+1}_{n}$, where n is the number of sides of the squares around the boundary of the rectangle.
Counting the sides of the unit squares in the grid that adjoin one another to form two adjacent sides of the rectangle and then doubling the count.	$P = (h + b) + (h + b)$, where h is the number of sides of squares on one side of the rectangle (its height), and b is the number of sides of squares on the adjacent side of the rectangle (its base).
Counting the sides of the unit squares in the grid that adjoin one another to make up the height of the rectangle, then counting the sides that adjoin one another to make up its base, then adding the two counts, and finally multiplying this sum by 2.	$P = 2(h + b)$, where h is the height and b is the base of the rectangle.
Counting the sides of the unit squares in the grid that adjoin one another to form each of the four sides of the rectangle and then adding these four side lengths.	$P = s_1 + s_2 + s_3 + s_4$, where s_x is the length of each side of the rectangle.
Counting the sides of the unit squares in the grid that adjoin one another to make up the height of the rectangle and multiplying this count by 2, then counting the sides that adjoin one another to make up the base of the rectangle and multiplying that count by 2, and finally adding the two products.	$P = 2h + 2b$, where h is the height, and b is the base.

Fig. 1.11. Counting methods and associated symbolic representations for determining the perimeter of a rectangle drawn on a square grid of unit squares

Linking the symbolic representation in this way with the physical act of counting to determine perimeter can help students understand why a specific formula or generalization "works." It can give them a deeper understanding of the relationships expressed in a formula, positioning them to apply the formula more appropriately. It helps students interpret their results and determine the reasonableness of their solutions. Additionally, explicitly discussing the multiple techniques for finding perimeter and linking them to the associated symbolic representation gives students multiple strategies to use, helping ensure that if they forget one strategy, they have another one available.

Attending to justification (Big Idea 4)

As emphasized in Big Idea 4, conjecturing and justifying conjectures are inextricably bound up with classifying, naming, defining, and posing—all the activities that intertwine in geometric investigation (Sinclair, Pimm, and Skelin 2012). Reasoning and proof serve several critical purposes in the geometry classroom. At the middle school level, instruction that nurtures students' ability to make conjectures, justify, verify, and explain is essential. You should encourage your students to discuss conjectures—their own and those made by others—and offer justifications for them. Ensuring that they have opportunities to critique the reasoning behind conjectures is part of building a classroom climate that establishes both respect and rigor. Actively and regularly inviting students to make conjectures and justify them are pedagogical practices that are tightly aligned with NCTM's Geometry Standard, which expects instruction in the middle grades to enable students to "create and critique inductive and deductive arguments" (NCTM 2000, p. 232), while it engages them in "making and validating conjectures, and classifying and defining geometric objects" (NCTM 2000, p. 233).

To help students to provide clear justifications for their mathematical conjectures and to support their efforts to shape those justifications into mathematical arguments (including foundational components of theory and proof that they will use in subsequent grades), instruction must move the students away from seeing mathematics as a set of irrefutable rules. Instead, students need to evaluate conjectures and statements and apply the generalizations that they make to other contexts so that they can probe them more deeply.

This pedagogical goal can be accomplished in multiple ways, but a method that you can use to engage your students actively and effectively in the work of examining, explaining, and justifying—or refuting—conjectures is to introduce the work of fictitious, "other" students for their consideration. To do this, you should generate statements to present to your students as the work of "a student in another class" or "other groups of students," and have your students evaluate their accuracy. The statements could include a variety of assertions; below are just two examples:

- "As the perimeter of a rectangle increases, the area also increases."
- "The units used to measure area must be identical in shape to the shape measured—squares must be used for squares, circles for circles, and so on."

Then you should pose such questions as the following:

- "Why did the group make that statement?"

- "How would you respond to their solution?"

- "Can you explain what they are thinking?"

Consistent use of the practice of having students explain the ideas of others can reinforce their awareness of the need for self-examination of their own reasoning and logic.

Use of this teaching practice gives students an opportunity to apply their own conjectures in the process of critiquing the reasoning of other students. It can extend and challenge students' thinking about the relationship between perimeter and area in a different context while giving you meaningful opportunities to assess their understanding.

Extending and assessing students' understanding of perimeter and area relationships

The students whose work on tasks A and B was discussed previously made substantive observations and conjectures regarding the relationships between perimeter and area (see figs. 1.4–1.9). Nevertheless, to internalize these ideas and make them part of an enduring, growing mathematical understanding, students must encounter them and apply them in more than one task. The major points about perimeter and area relationships seem relatively straightforward in tasks A and B. However, calling these points to mind and putting them to work in a new setting are challenging, especially when students are asked to measure shapes that do not offer the uniformity of rectangles or other regular figures. Figure 1.12 shows the Leaf task (Ronau and Gilbert [1988]; adapted by Wilson and Chavarria [1993]), designed to extend and challenge students' thinking about perimeter and area. Examine this task, guided by the questions in Reflect 1.6.

Reflect 1.6

Figure 1.12 presents the Leaf task (adapted by Wilson and Chavarria [1993] from Ronau and Gilbert [1988]).

How does the structure of this task compare with that of the paired tasks, A and B, in figure 1.4?

Suppose that you gave this task to your students. What misconceptions do you think their work on the task might elicit, bringing those misunderstandings to the surface for the students' own examination as well as your assessment?

Russell's group and Jordan's group were asked to find the area of the leaf shown below. Decide whether each group used accurate methods, and explain your thinking.

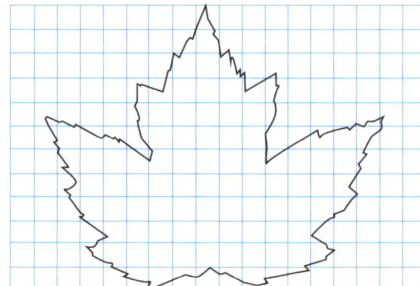

- Russell's group found the area of the leaf by counting all the squares inside the leaf's boundary. His group paired half squares to create full square units and totaled all the square units.

- Jordan's group took a string and placed it closely around the perimeter of the leaf. Then they created a rectangle with the measured string and counted the length and width of the shape in sides of squares and multiplied them to find the area.

Fig. 1.12. The Leaf task, a perimeter and area problem using a nonuniform shape. Adapted by Wilson and Chavarria (1993) from Ronau and Gilbert (1988).

The Leaf task presents an irregular shape drawn on a grid, along with solutions purportedly from two groups of students—Russell's group and Jordan's group. Students completing the task are asked first to decide whether they would expect the methods that the groups used to produce an accurate result and then justify their determination. The methods that Russell's and Jordan's groups used are likely to be similar to those used by students to solve tasks A and B, but the groups are now applying them to a shape that does not have an associated formula. In evaluating the solution strategies offered by Russell's and Jordan's groups, students must draw on their own understanding of the concepts of perimeter and area and the relationships between them.

We gave the Leaf task to the same group of sixth-grade students that we had previously asked to examine tasks A and B (see fig. 1.4)—the group that produced the five samples of student work shown in figures 1.5–1.9. Figures 1.13–1.15 show responses to the Leaf task from three of the five sixth graders whose successful work on the more straightforward task is included in the previous group of figures. Ji's work on the Leaf task appears in figure 1.13 (fig 1.9 shows Ji's earlier work),

Karla's work appears in figure 1.14 (fig. 1.5 shows her earlier work), and Kris's work appears in figure 1.15 (fig. 1.7 shows Kris's earlier work). Their conclusions about the accuracy of Russell's and Jordan's groups' solution methods and their justifications of their determinations indicate their varying levels of understanding of the relationships between perimeter and area.

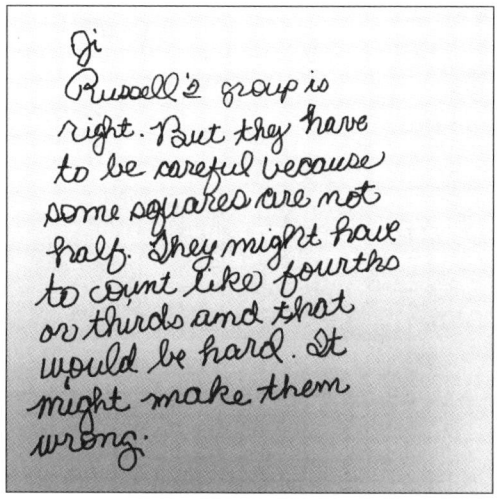

Fig. 1.13. Ji's (grade 6) response to the Leaf task

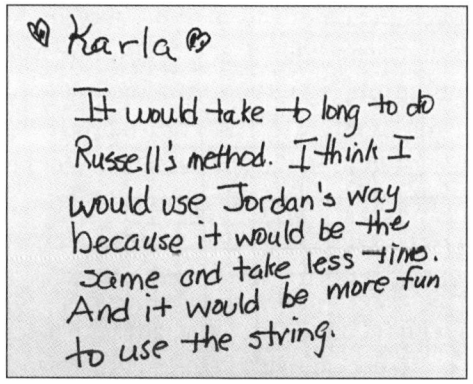

Fig. 1.14. Karla's (grade 6) response to the Leaf task

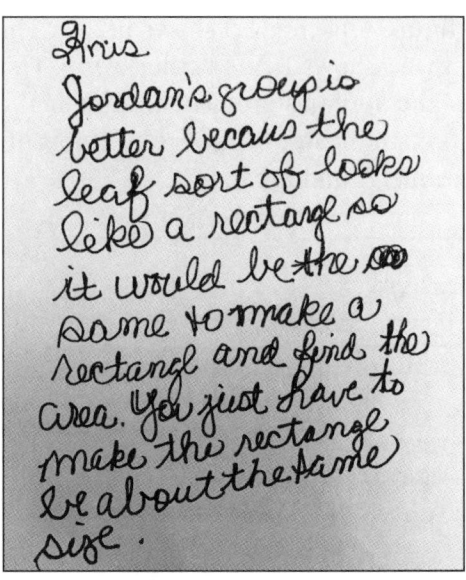

Fig. 1.15. Kris's (grade 6) response to the Leaf task

Note the high value that Karla placed on efficiency relative to accuracy in her assessment of the groups' methods. Her response seems to suggest that she believed that both methods could yield an accurate result, but "it would take to [sic] long to do Russell's method." Students may suggest that the string approach used by Jordan's group is superior to the counting method used by Russell's group for a variety of reasons that either ignore accuracy or assume it for both methods. Their reasons in favor of Jordan's group's method may include more efficient use of time or greater ease of computing or counting. According to Kris, by applying Jordan's "better" method, "You just have to make the rectangl [sic] be about the same size [as the leaf]." Ji was the only student of the three who correctly stated that the counting method used by Russell's group was "right." Nevertheless, Ji pointed to the inefficiency and potential for error in the method:

> But they have to be careful because some squares are not half. They might have to count like fourths or thirds and that would be hard. It might make them wrong.

Students' assessment of this drawback in Russell's group's method may go beyond Ji's. Instead of simply urging care with Russell's counting method, they may conclude that making complete squares as the method calls for from such "messy" pieces along the edge of the leaf is so difficult as to make the method impractical to use or even unlikely to yield, or incapable of yielding, an accurate result. By contrast, because the string method used by Jordan's group beguilingly offers a way to create a "nice" rectangle and facilitates performing a deceptively easy calculation of the area through multiplication, students may identify it as the "better" method.

Clearly, in spite of their observations about perimeter and area relationships in tasks A and B, students tend to be drawn to a method that contradicts the very observations that they made earlier. If, after responding to tasks A and B, your students evaluated Russell's and Jordan's groups' methods and decided that Jordan's group's method was "better," how could you challenge this conclusion and motivate them to reflect on and reassess their thinking? One possibility would be to project the leaf on the whiteboard and use string to measure around its edge. You could then cut the string to give a length equal to the perimeter and tape the ends together to form a loop, which you could show to your students and ask, "If this string is the same length as the perimeter of the leaf, then should it enclose the same area as the leaf?" From those who believe that it should, you could select a student to create a rectangle whose area he or she thought was similar to that of the leaf. Then you could take the loop of string and, by pulling it into a narrow rectangle with very little area, show a counterexample. Some students would almost certainly be surprised, at which point you could ask them to reflect on the observations that they previously made about tasks A and B. This process may motivate them to question why they did not use their thinking from those tasks—specifically, their conclusion that the area cannot be known from the perimeter—to make their decision in the Leaf task.

The fact is that the Leaf task is very challenging—particularly because it offers a measuring approach that students find very inviting, lending itself as it does to efficient use of a familiar procedure and formula. Even though students may have had previous experiences in which they produced multiple figures with fixed perimeters and areas and developed generalizations about the relationship between perimeter and area, they are often unable to transfer that knowledge to an unusual application such as the Leaf task, which calls on them to move beyond a procedural approach. However, confronting students with a task such as this one after they have investigated the previous, more straightforward tasks A and B gives them an important opportunity in a different, more complex context to reinforce their previous observations and reflect that procedural knowledge regarding the use of formulas may not be enough and that building on and applying earlier discoveries and understandings is critical.

Extending understanding to three-dimensional measures

CCSSM first addresses three-dimensional geometric measurement in grade 5, where students are expected to explore "and understand concepts of volume and relate volume to multiplication and addition" (NGA Center and CCSSO 2010, p. 37 [5.MD]). CCSSM expects fifth graders to use unit cubes to pack a rectangular prism without gaps or overlaps to obtain a volume measurement, then progress to the use of standard and improvised cubic units to measure volume, and ultimately develop the formulas $V = l \times w \times h$ and $V = b \times h$ for the volume of a rectangular prism (5MD.3–5).

This initial development of students' understanding of volume at the end of the upper-elementary grades provides the foundation for the gradual expansion of concepts and formulas for other, more complex three-dimensional shapes, such as cones and spheres, in grades 6–8. This middle-grades expansion is elaborated in NCTM's Geometry Standard and CCSSM, both of which expect students to move away from hands-on experiences in packing rectangular prisms with cubes and counting them to more abstract visualizations of three-dimensional shapes in two-dimensional space, with increasing attention to relationships between and general-izations about surface area and volume. As NCTM's Geometry Standard expresses it, students in grades 6–8 should be able to "use two-dimensional representations of three-dimensional objects to visualize and solve problems such as those involv-ing surface area and volume" and should "use visual tools such as networks to represent and solve problems" (NCTM 2000, p. 232). CCSSM traces the arc of this work from grade 6 to grade 8. In grade 6, CCSSM expects students to build on their initial work in grade 5 by using unit cubes of the appropriate unit fraction edge lengths to pack a right rectangular prism with fractional edge lengths (6.G.2). Then they should show that the volume that they measure in this manner is the same as that obtained by multiplying the prism's edge lengths (6.G.2). This work leads, in grade 7, to solving problems "involving area, volume, and surface area of two-and three-dimensional objects" (7.G.6), and ultimately, in grade 8, to students' develop-ment of the formulas for the volumes of cones, cylinders, and spheres and the use of these formulas to solve real-world and mathematical problems (8.G.9).

Surface-area concepts and skills follow directly from students' understanding of the area of two-dimensional shapes as well as from their understanding of nets. Nets—two-dimensional representations of three-dimensional shapes—provide opportunities for substantive spatial explorations as well as opportunities to link a physical model to surface-area formulas. Consider, for example, the task in figure 1.16 while responding to the questions in Reflect 1.7.

Reflect 1.7

Figure 1.16 shows a task that asks students to find all the unique nets of a cube. Suppose that you gave your students this task.

What ideas about a cube do you think their exploration of all the different nets might support?

What ideas about three-dimensional shapes in general might their work help to reinforce?

Below is a net for a cube:

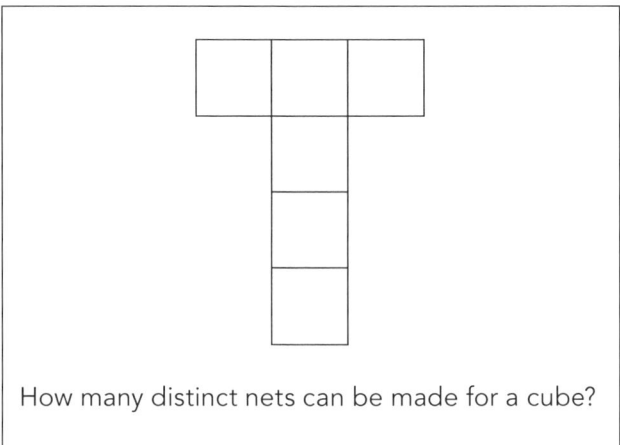

How many distinct nets can be made for a cube?

Fig. 1.16. A task asking students to find all the nets for a cube

Eleven nets can be made for a cube, as shown in figure 1.17. In each of these eleven nets, students can easily see that a cube's faces are six squares. Students' work in finding the nets and examining them can help them see that the surface area of the cube is the sum of the areas of the squares, or six times the area of one of the squares. Similar explorations with nets of other three-dimensional shapes can result in the development of a better understanding of surface area.

Figure 1.17. The eleven unique nets of a cube

Grasping the concept of the surface area of a three-dimensional shape requires the application of ideas about the area of a two-dimensional shape and recognizing the direct link between the two. A net physically models area, showing all the faces of a three-dimensional shape and making the calculation of the shape's surface area straightforward and obvious.

However, finding volume is not quite as obvious. Just as computing area requires students to coordinate two dimensions—and two length measures—to produce a two-dimensional measurement, applying the volume formula requires students to coordinate three length measures to produce a three-dimensional measurement. For some students, this process is not intuitive, and their difficulty in conceptualizing it often results in their applying formulas for volume in a procedural fashion rather than with understanding.

Battista (2003) describes four processes that students must use to become competent in working with volume, as detailed in figure 1.18. These include (1) forming and using mental models, (2) attending to spatial structuring, (3) engaging in what Battista calls "units locating," and (4) organizing by composites. Students' internalization and mastery of these four processes rely on their having the many and varied experiences with physical models that they need to develop them fully.

Process	Example
Forming and using mental models	Visualizing and reasoning about situations
Spatial structuring	Abstractly identifying, interrelating, and organizing objects' components
Units locating	Mentally locating a cube, for example, within a three-dimensional object
Organizing by composites	Creating composite units from a cube to form a new unit

Figure 1.18. Battista's four mental processes for competence in working with volume (2003)

Battista's four processes are a foundation for subsequent, more sophisticated thinking and visualizing related to volume. Having multiple experiences in measuring volume enables students to develop and use these processes fluently, and this competency supports them in achieving the larger, more general, more abstract insights that Sinclair, Pimm, and Skelin (2012, p. 8) identify as Big Ideas 2 and 3, which address work with imagery and mental objects:

Big Idea 2
Geometric thinking involves developing, attending to, and learning how to work with imagery.

Big Idea 3
A geometric object is a mental object that, when constructed, carries with it traces of the tool or tools by which was constructed.

Essential Understanding 2c, one of the three subordinate ideas that Sinclair, Pimm, and Skelin (2012) associate with Big Idea 2, resonates particularly with the four processes that Battista asserts that students must use with competence with volume: "Geometric awareness develops through practice in visualizing, diagramming, and constructing" (p. 8).

Many tasks paralleling the perimeter and area tasks presented in this chapter offer opportunities to expand students' thinking and understanding of measures related to 3-dimensional shapes (surface area and volume). Higher-order questions can push students' thinking, motivating them to create conjectures, test them, and use them to justify other conjectures or observations. Some tasks to consider might include the following:

1. Ben said, "As the volume of a prism increases, the surface area of the prism also increases." Do you agree with Ben? Why or why not? Support your answer.

2. The edge of a cube measures 5 cm. Another cube has an edge length of 10 cm. Without performing any calculations, make a conjecture about the relationship between the volumes of the two cubes. Also make a conjecture about the relationship between the surface areas of the two cubes. Justify your answers.

3. Below are two-dimensional representations of an open-top box:

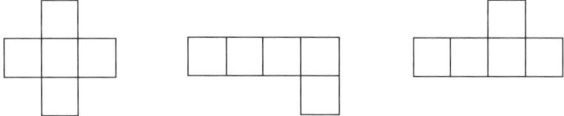

   a. How many ways can you add a square to create a net for a cube?

   b. Find three other two-dimensional representations that will fold into an open-top box.

    c. How many ways can you add a square to your other representations to create a net for a cube?

    d. What observations can you make?

Additional tasks are identified in Appendix 2; see especially the descriptions of articles by Garimella and Robinson (2015); Haberern (2016); Jeon (2009); and Prummer, Amador, and Wallin (2016). Implementing tasks beyond traditional computation problems, which require only an application of a formula, provides opportunities to extend students' thinking. As students apply conjectures and generalizations that originated in their own experiences, they gain a deeper understanding of relationships within these measurement contexts.

Summarizing Pedagogical Content Knowledge to Support Big Idea 1, Essential Understanding 1*a*, and Big Idea 4

Teaching the mathematical ideas in this chapter requires specialized knowledge related to the four components presented in the Introduction: learners, curriculum, instructional strategies, and assessment. The four sections that follow summarize some examples of these specialized knowledge bases in relation to the geometric measurement in connection with Big Idea 1, Essential Understanding 1*a*, and Big Idea 4.

Knowledge of learners

Studies have shown that students' geometric understanding correlates with their spatial visualization and achievement in other areas of mathematics that may not appear to be geometric (de Hevia, Vallar, and Girelli 2008; Stavridou and Kakana 2008). In fact, strong spatial visualization skills have been shown to predict student success in the areas of science, technology, engineering, and mathematics (for example, see Shea, Lubinski, and Benbow [2001] and Wai, Lubinski, and Benbow [2009]).

However, students' experiences from previous grades in geometry may have been less than optimal in developing their understanding, for a variety of reasons. For example, geometric concepts and skills are often taught at the end of the academic year to allow more time to teach number topics. CCSSM includes fewer standards for geometry than for number in the elementary grades, and consequently teachers may perceive geometry as an area that merits less emphasis than the number domains.

Therefore, students entering middle school may be functioning at lower levels of thinking on the van Hiele continuum (Battista 2002; Senk 1989). The van Hiele model includes five levels of thinking, usually numbered 0–4:

- Level 0, Visualization

- Level 1, Analysis

- Level 2, Informal deduction

- Level 3, Formal deduction

- Level 4, Rigor

Thinking at each level is progressively more sophisticated than at the previous level, culminating in thinking at level 4, rigor, at which point students are capable of operating within an axiomatic system and making relatively abstract deductions. The goal in grades 6–8 is to move students at least to level 2, informal deduction, at which point they can identify and work with relationships among shapes and their properties.

However, students arriving at the middle grades may be functioning only at level 0, visualization, where they are limited to naming shapes. They may be able to name some properties of the shapes but lack confidence about the characteristics that are necessary or sufficient for those shapes. If they have not developed beyond visualization, the way in which they approach a task is likely to be to focus on its visual aspects. This can result in immature or overgeneralized observations or rules. For example, they may leap to the generalization that as the perimeter of a rectangle increases, the area of the rectangle also increases.

It is thus important to recognize that students may bring to the classroom a diverse set of experiences. More instructional time may be needed so that students have significant experiences that can move their thinking from a rudimentary level to a more sophisticated level that will allow them to consider relationships among geometric shapes and their properties thoughtfully and analytically.

Knowledge of curriculum

Knowing that students come into middle grades with varied experiences in geometry can affect and potentially strengthen your selection of tasks for use in your instructional sequence. Tasks of types that you might otherwise have chosen may seem less appropriate than others in light of this realization. The tasks presented in this chapter offer students opportunities not only to explore perimeter and area but also to confront misconceptions that they may have formed about these foundational concepts. These new encounters enable students to revisit notions that they have imagined to be true in the context of tasks that challenge their thinking. The resulting cognitive dissonance motivates them to resolve the contradictions between previously held ideas and the mathematics that they are seeing, using, and thinking about in the task that they are currently doing.

Tasks that ask students to create multiple solutions function as generalization tasks. As students share their solutions, they have opportunities to observe common characteristics that then lead them to generalizations. For example, task A (fig. 1.4), which asks students to use square units to create as many rectangles as possible with an area of 36 square units, provides them with an opportunity to observe that a rectangle with a fixed area can have different dimensions and thus does not, unless otherwise specified, have a fixed perimeter. In fact, when students use unit squares to create rectangles with an area of 36 square units, they discover that they can create five distinct rectangles with very different dimensions and perimeters: 74 units (1 unit by 36 units), 40 units (2 units by 18 units), 30 units (3 units by 12 units), 26 units (4 units by 9 units), and 24 units (6 units by 6 units). Likewise, task B (fig. 1.4), which asks students to use square units to create as many rectangles as possible with a perimeter of 24 units gives them a chance to observe that a rectangle with a fixed perimeter can also have different dimensions and thus does not have a fixed area. The students' work on the task allows them to discover that, by using square tiles, they can create six distinct rectangles with perimeters of 24 units but with very different areas: 11 square units (1 unit by 11 units), 20 square units (2 units by 10 units), 27 square units (3 units by 9 units), 32 square units (4 units by 8 units), 35 square units (5 units by 7 units), and 36 square units (6 units by 6 units).

Both the richness of the tasks and the sequencing of them in the classroom are critical to the success of these kinds of learning opportunities for students. One fairly typical characteristic of a rich task is a call for students to create multiple cases, as in "create as many different rectangles as possible, each with an area of 36 square units" (task A) or "create as many different rectangles as possible, each with a perimeter of 24 units" (task B). Creating these multiple cases allows students to take note of and internalize relationships between perimeter and area. Specifically, they should note that knowing the perimeter of a rectangle does not mean knowing its area, and, conversely, knowing the area of a rectangle does not mean knowing its perimeter. These insights then lead to a realization that an increase in the area of a rectangle does not necessitate an increase in its perimeter. Notice that these tasks are neither complex nor difficult; they are accessible to all students, but the solutions to the tasks provide opportunities to move the simple task to a much deeper level.

The sequencing of the tasks also is linked to their richness. In using rich tasks, teachers do not begin with skill development. Initial tasks do not focus on the formulas for geometric measures. Instead, early tasks begin with the conceptual aspects related to perimeter and area, allowing students to explore geometric measurement in accessible tasks. Later lessons, beyond the scope of this chapter, can focus on the skill of using a formula for finding area and perimeter.

Knowledge of instructional strategies

As Young (1911) remarked more than a century ago, geometry instruction has often included tasks that focus on memorization of definitions and procedures. This chapter, however, has presented tasks that emphasize student exploration and sharing of both solutions and solution strategies. Tasks such as the Leaf task encourage students to share their thinking in a very natural way and critique the thinking or reasoning of others. Incorporating tasks such as these requires that you use instructional strategies effectively to optimize the learning outcomes.

Lamberg (2012) provides strategies that can be used to facilitate class discussions effectively. She notes that small-group and whole-class discussions can lead to (1) a common or shared understanding of a problem and its solution, (2) the development of students' metacognition through teacher questioning, and (3) opportunities for students to evaluate and analyze their own and their peers' reasoning. According to Lamberg, the facilitation of the discussions should then be thought of as having three phases:

> Phase 1: Making thinking explicit
>
> Phase 2: Analyzing solutions
>
> Phase 3: Developing new mathematical insights

In each phase, teacher questioning provides the impetus for the discussion. However, as students become accustomed to a discourse-based classroom, their questions also move the discussion forward because they begin to emulate the teacher's questioning. Those questions will then lead to their own questions.

Smith and Stein (2011) identify five practices that teachers can use to promote productive discussions. These practices are as follows:

- Anticipating: Solve the problem yourself beforehand, thinking about how your students are likely to solve it and determining the mathematics that is embedded in it.

- Monitoring: Listen and observe students while they solve the problem, identifying strategies that they are using and refocusing them through questions if they are not on track.

- Selecting: Decide what solution approaches or solutions you want to highlight, identifying ones that will further the mathematical thinking.

- Sequencing: Identify the order in which you want students to share their thinking or solutions, determining the method by which students will share.

- Connecting: Ask questions that connect the solutions and the mathematical ideas to highlight, comparing multiple students' solutions and methods.

While students are sharing, they will be thinking critically about the solutions and the solution methods. You will also be analyzing the students' work, using their solutions as a way to assess the level of their thinking and the depth of their understanding.

Knowledge of assessment

The tasks presented in Chapter 1 provide opportunities for you to assess student learning through their sharing of solutions and discussions. According to Wiliam (2007), tasks that provide such opportunities can function as types of formative assessments, enabling teachers to make instructional decisions that are based on the understandings and misunderstandings that come to the surface in the students' work and their discussions of it. You can gather information about their thinking, both while you circulate as they work on the task and while you attend to their approaches and solutions in whole-class discussions after they have completed the task. The use of reversibility, flexibility, and generalization tasks, as described by Dougherty (2001) and illustrated in the Introduction, enhance the information gained from student solutions.

By presenting tasks that prompt students to solve in multiple ways or find multiple solutions, you can determine the depth of their understanding with greater precision. The questions that you can then create to extend the problem or explore or probe student thinking can provide you with further evidence of the way in which students have constructed their knowledge about geometric measurement. These windows on their understanding can also show you limitations of their thinking that may hinder their ability to use particular approaches to solving the tasks.

It is worth emphasizing that you need to determine what your students *don't know,* as well as what they do know. In addition to giving you insight into the depth of their understanding, tasks of the kind presented in Chapter 1 allow you to discern pervasive student misconceptions that need further explorations. Explicitly addressing misconceptions by providing tasks that create cognitive dissonance or that make a misconception evident is critically important for advancing students' understanding.

Conclusion

Chapter 1 has demonstrated possibilities for designing instruction that can provide optimal opportunities for students to develop an understanding of geometric measurement that goes beyond rote memorization. Effective instruction requires that you understand your students as learners, identify and implement rich tasks, and use student responses to the tasks to make instructional decisions. The tasks presented in this chapter have provided some examples of tasks that can support your teaching of geometric measurement.

Chapter 2 continues to highlight big ideas and essential understandings from *Developing Essential Understanding of Geometry for Teaching Mathematics in Grades 6–8* (Sinclair, Pimm, and Skelin 2012). Another major strand of middle-grades geometry becomes the focus in that chapter: transformational geometry.

practice

Chapter 2
Transformations

Big Idea 3
A geometric object is a mental object that, when constructed, carries with it traces of the tool or tools by which it was constructed.

Essential Understanding 3*b*
Geometric thinking turns tools into objects, and in geometry the process of turning an action undertaken with a tool into an object happens over and over again.

Geometry focuses on mental objects—pure abstractions. However, these mental objects are closely connected with the physical world and the tools and processes that we use with them. This is the central insight that Sinclair, Pimm, and Skelin explore as Big Idea 3 in *Developing Essential Understanding of Geometry for Teaching Mathematics in Grades 6–8* (2012). As they explain,

> Geometric objects are not everyday objects (like the lids of cans or the bases of cups) but are, in part, derived from them and then extended or otherwise altered mentally. ... Although we are accustomed to seeing clean diagrams of circles on the printed page, we often forget how much the idea of a circle is intertwined with the tools used to create it. (pp. 40–41)

In other words, although the objects of study in geometry are abstractions—mental objects that do not exist in the physical world—our knowledge of them is rooted in and is shaped by the physical world and the concrete tools and the processes that we use when we work with them and think about them.

The essential understandings associated with Big Idea 3 are aspects of this over-arching insight. Tools and processes "enable us to create new geometric objects" (Sinclair, Pimm, and Skelin 2012, p. 44), a perception that leads to Essential Understanding 3a: "Tools provide new sources of imagery as well as specific ways of thinking about geometric objects and processes" (p. 41). To illustrate this idea, Sinclair, Pimm, and Skelin use the example of rotating an object around a point—perhaps by using a compass as a tool, with one leg anchored at the point and the other attached to a pencil that traces the path of the rotating object. This process of *rotating* leads to the abstract idea of a *rotation*—a mental object "that has properties of its own" (p. 44). In fact, as Sinclair, Pimm, and Skelin remark, "Many ... tools have been and can be used as sources of mental imagery and objects" (p. 40). Using a mirror or folded paper to flip, or *reflect*, an object leads to the idea of a *reflection;* physically sliding, or *translating*, an object from one place on a surface to another spot leads to the idea of a *translation*; and so on. In fact, the frequency with which this experience repeats itself in geometry leads to the insight that Sinclair, Pimm, and Skelin present as Essential Understanding 3b: The experience of turning processes or tools into mental objects for further study happens again and again in geometry. In this way, Sinclair, Pimm, and Skelin underscore the tight connection between Big Idea 3 and its essential understandings—particularly, Essential Understanding 3b—and transformations, which are an important focus of geometry in the middle grades and the focus of this chapter.

A transformational approach to geometry offers rich contexts for discussing various topics by allowing middle-grades students to use their intuitive understanding of objects and motion (Jones 2000; Martin 1996). For middle-grades students, this approach can be very productive, uncovering, highlighting, and reinforcing connections among concepts such as measurement, fractions, ratios, percentages, and proportional reasoning (Slovin 2000). Furthermore, a focus on transformations allows middle-grades students to explore a central idea that bridges many mathematical topics: the concept of invariance (Jones 2000).

The importance of focusing on transformations in the middle grades is underscored in both the NCTM Standards (NCTM 2000) and the Common Core State Standards for Mathematics (CCSSM; National Governors Association Center for Best Practices and Council of Chief State School Officers [NGA Center and CCSSO] 2010). The NCTM Geometry Standard expects students in grades 6 to 8 to—

- describe sizes, positions, and orientations of shapes under informal transformations such as flips, turns, slides, and scaling; [and]

- examine the congruence, similarity, and line or rotational symmetry of objects using transformations. (NCTM 2000, p. 232)

Likewise, CCSSM expects middle-grades students to demonstrate proficiency in these tasks by the end of grade 8. In grade 6, CCSSM recommends that students lay the groundwork for exploring dilations. Sixth graders are expected to "prepare for work on scale drawings and constructions in Grade 7 by drawing polygons in the coordinate plane" (NGA Center and CCSSO 2010, p. 40). CCSSM expects students in grade 7 to encounter the idea of a dilation very directly as they—

> Solve problems involving scale drawings of geometric figures, including computing actual lengths and areas from a scale drawing and reproducing a scale drawing at a different scale. (NGA Center and CCSSO 2010, p. 49 [7.G.1])

After this work with similarities in grade 7, students are expected to add isometries in grade 8 and use their work with transformations to build concepts related to congruency and similarity. CCSSM elaborates five standards for grade 8 tying work with transformations to an understanding of congruence and similarity (NGA Center and CCSSO 2010, pp. 55–56 [8.G.1–5]):

Understand congruence and similarity using physical models, transparencies, or geometry software.

1. Verify experimentally the properties of rotations, reflections, and translations:

   a. Lines are taken to lines, and line segments to line segments of the same length.

   b. Angles are taken to angles of the same measure.

   c. Parallel lines are taken to parallel lines.

2. Understand that a two-dimensional figure is congruent to another if the second can be obtained from the first by a sequence of rotations, reflections, and translations; given two congruent figures, describe a sequence that exhibits the congruence between them.

3. Describe the effect of dilations, translations, rotations, and reflections on two-dimensional figures using coordinates.

4. Understand that a two-dimensional figure is similar to another if the second can be obtained from the first by a sequence of rotations, reflections, translations, and dilations; given two similar two-dimensional figures, describe a sequence that exhibits the similarity between them.

5. Use informal arguments to establish facts about the angle sum and exterior angle of triangles, about the angles created when parallel lines are cut by a transversal, and the angle-angle criterion for similarity of triangles. *For example, arrange three copies of the same triangle so that the sum of the three angles appears to form a line, and give an argument in terms of transversals why this is so.*

Rigid transformations, such as translations, reflections, and rotations, maintain congruence and similarity; that is, the shape and size of the object are invariant under the transformation. Figure 2.1, for example, shows how the parallelogram *ABCD* can be visualized as two congruent triangles, $\triangle ABD$ and its rotated image, $\triangle CDB$. In contrast, similarity, but not necessarily congruence, is invariant under dilations; that is, a dilation and its pre-image are similar, having the same shape, but are not necessarily congruent, because they might not be the same size.

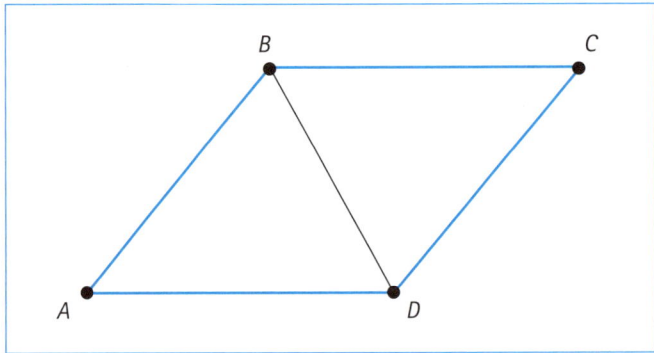

Fig. 2.1. Parallelogram *ABCD* as the combination of two congruent triangles

The idea of the slope of a straight line, a central concept for middle-grades students, can be developed by using the notion of similarity. Figure 2.2 illustrates how arbitrary similar right triangles *ABC* and *ADE* can be used to visualize the slope of a line containing the points *A*, *B*, and *D*. Furthermore, the slope of a line is constant, since

$$\frac{BC}{AC} = \frac{DE}{AE}.$$

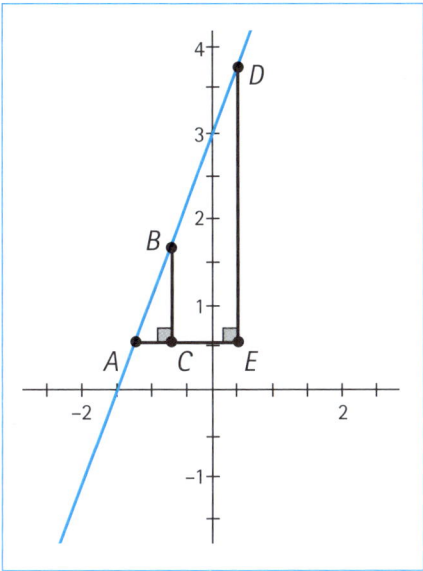

Fig. 2.2. Two similar triangles, each representing the slope of a line

Using a geometric rather than a numerical approach to slope may offer new opportunities to students who have struggled with a heavily number-based curriculum. A geometric approach may help them gain access to significant mathematics content that previously seemed obscure to them. In addition to providing students with a new perspective from which to work with important mathematical ideas, the CCSSM grade 8 geometry standards take an active approach to building concepts about transformations. And in keeping with this approach, it is worth remembering that, as Sinclair, Pimm, and Skelin (2012) emphasize, "The central advantage of working with transformations is to maintain the connection between the processes and the objects of geometry" (p. 54).

Working toward Big Idea 3 and Essential Understanding 3*b*

When we talk about geometric transformations, we are referring to a one-to-one correspondence that pairs each point of the plane with another (not necessarily different) point of the plane. By convention, the first point is often called the *pre-image* and the second point is called the *image*. *Isometry* means "equal measure" or "the same measure" in Greek, and transformations that preserve distances are called *isometries*. Isometries model motions that send "lines to lines and preserve many properties of the shapes, including length, angle measure, and area" (Sinclair, Pimm, and Skelin 2012, p. 44), or, said another way, these properties are *invariant* under isometries. The mapping, or fitting, of one shape on top of another is one way to verify congruency of the shapes and is a strategy that can be traced back at least to Euclid and his use of the method of superposition (Kline 1972, p. 87). However, this heavy focus on the shapes, or objects, can create a misconception. It can make understanding transformations as mathematical models of motions of the *entire* plane difficult for students, who often think of the plane as empty space (Edwards 1991) and commonly view the transformation as happening only on, or to, the object or objects of interest.

Research, including the work of Küchemann (1981) and Masters (2010), for example, has found that the failure to understand that a transformation applies to the entire plane is sometimes problematic. Tasks in which the line of reflection is perpendicular to an edge of the page and the object of interest is close to and aligns with the line of reflection do not always cause difficulties for students. Understanding that a reflection is a transformation of the entire plane becomes more critical, however, when the line of reflection (or center of rotation) is at a distance from the object or when the line is oblique (Masters 2010). This misunderstanding about transformations is often compounded by the language used in the classroom. Teachers or curriculum may instruct students to reflect or rotate a given object, implying that the transformation applies only to that object and not the entire plane.

Students may benefit from experiences that emphasize that all the important characteristics are contained in the transformation itself and its own properties, irrespective of any particular object or objects. Students can in fact completely understand a transformation without being given a specific accompanying object of interest. They can learn that a reflection is determined by a given line; a rotation, by a point and an angle; a translation, by a vector; and a dilation, by a point and a scaling factor. A particular object may have special interest (for example, "What is the image of the parallelogram *ABCD*?"), or it may be given as a visual aid. However, students must understand that a transformation affects the whole plane, not just specific objects of local interest.

Sequencing problems to build understanding of reflections

In middle-grades work with reflections, as with all isometries, students need experience with problems crafted in a way that highlights the transformation of the plane. Using physical tools and working concretely can support this fundamental understanding. Students benefit from using transparencies or tracing paper, a Mira (see fig. 2.3), and other physical models, along with digital tools, to experiment with solutions to problems. Such work will engage them in the core mathematical practice that CCSSM's Standards for Mathematical Practice present as Mathematical Practice 5: "Use appropriate tools strategically" (NGA Center and CCSSO 2010, p. 7). Building on their insights and discoveries in these experiments, students begin to form the informal arguments that provide the rationale for what will later become formal proofs, thus supporting them in engaging in another core mathematical process, identified as Mathematical Practice 2 in CCSSM: "Reason abstractly and quantitatively" (NGA Center and CCSSO 2010, p. 6).

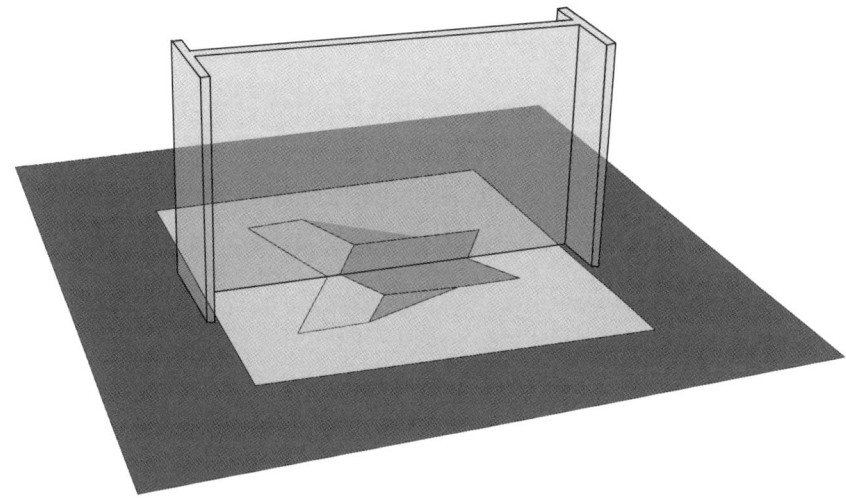

Fig. 2.3. A Mira used to create the reflected image of an object. From Sinclair, Pimm, and Skelin (2012), p. 45.

A sequence of four problems follows to illustrate one way to introduce reflections, move students forward, and assess their understanding. These problems, which offer increasing challenge as student progress through them, are designed to give students both "local" and "global" experiences. They enable students to see locally how the image of an object differs from but is related to the pre-image object, thereby enabling the students to experience the fundamental properties of a reflection. In addition, the problems are designed to prompt a global discussion of the plane.

An important feature of the four problems in the sequence is that they are designed to develop students' understanding of the concepts related to reflections progressively, over the sequence, instead of attempting to embody all the key concepts in a single task and address all of them at once. The suggestions for implementing the problems support an approach to teaching and learning in which students are able to build their understanding of reflections over a period of time.

Beginning with everyday reflection experiences

Effective initial tasks allow students to draw on and build on their previous experiences. Introductory problems in reflections should invite students to use familiar, everyday experiences with reflections in responding to the tasks. A well-designed introductory task will also give you an opportunity to assess the understanding that your students bring to the study of reflections, including any misconceptions that they may have. Consider the introductory problem in figure 2.4, guided by the questions in Reflect 2.1.

Reflect 2.1

Figure 2.4 shows an introductory reflection problem for middle-grades students.

What prior knowledge does this problem assume that students have?

Where do you think your students would place the bird's reflection in the lake?

What would you say to students who misplaced the reflection?

Kelly sees a bird flying over a still, smooth lake. She stands at the edge of the lake and also sees what looks like the same bird in the lake. Draw what Kelly sees in the water. Compare the bird with its reflection in the water. What is the same about the bird and its reflection? What is different about them?

Fig. 2.4. An introductory problem on reflections: Kelly sees a bird flying over a lake. From Slovin and colleagues (2003, p. 5).

The questions that accompany this problem are designed to lead students to begin identifying the properties of a reflection. Some students may confuse a reflection image with a shadow, especially since the context here is the sky, and, by implication, the sun. Asking students what is the difference between a shadow and a reflection and how they can tell which one they are seeing should lead the discussion to focus on the properties of a reflection. Students' drawings will vary. Some will create the bird's reflection with a freehand sketch, as in figure 2.5. Others will try some form of tracing. One common mistake is to trace the bird and rotate it 180 degrees, with the result that it ends up as shown as in figure 2.6.

Fig. 2.5. The bird and its reflection in the lake, with the reflection as a freehand sketch. From Slovin and colleagues (2003, p. 6).

You can help students assess the accuracy of their method by asking them what the image of the bird should look like. Your questions can direct their attention to various aspects of the reflected image, helping them think about different characteristics of a reflection. For example, your questions might include the following:

- Should the bird in the lake be the same size and shape as the bird in the air?

- Which way should the bird in the lake face?

- Where in the water should the bird in the lake be?

Fig. 2.6. The bird and its misplaced image in the lake, with the original bird rotated 180 degrees rather than reflected

This last question is important in drawing attention to the distance between the pre-image, the bird in the air in the given drawing, and the image, the bird as seen in the water. Questions such as these will not only acquaint your students with the properties of a reflection but also help them form the habit of sketching accurate drawings, a problem-solving process that is consistent with CCSSM's Mathematical Practice 6: "Attend to precision" (NGA Center and CCSSO 2010, p. 7).

Depending on how the given drawing is located on the student's paper and how much blank area surrounds it, the student may draw the image of the bird where he or she has room rather than where the image would be as determined by the distance between the pre-image and the line of reflection—in this case, the shore-line. The issue of available space on the paper presents an opportunity to raise the issue of the infinite plane and to suggest to students that when they are solving transformation problems, they may want to trace the given drawing on another sheet of paper so that they will have enough room—a representation of a large enough region of the plane—to locate the image accurately.

Although this sample initial problem uses a familiar example from everyday life—an outdoor scene with a reflection in water—its solution relies on students' ability to decontextualize the situation enough to envision a plane extending infinitely far in all directions, with only a small, finite region represented by the sheet of paper. The transition from working literally with contextualized referents to viewing these concrete elements more abstractly can be challenging for students, but the transition is important, since it allows them to extract the mathematics modeled by the example.

After solving and discussing this introductory problem, students should work with additional simple reflections and begin to define the properties of the motion. One readily available possibility is to have them look again at the problem of the bird flying above the lake and ask them to draw the reflection of the tree this time. This follow-up task will reveal their understanding of the surface of the lake as the line of reflection while also emphasizing the reflection of all the points in the plane.

If students are using tracing paper to create the images in their work with simple reflections, they can also begin to evaluate the precision of their drawings. They can consider issues such as the following:

- Are the pre-image and the image that they have drawn congruent?

- Do the points of their image coincide with the corresponding points of the pre-image when they fold the paper along the line of reflection?

- Have they labeled the objects in their drawings properly?

- After building an initial understanding of reflections through work with a number of everyday examples, students will be ready for a problem at a higher level of abstraction. Making predictions about reflected images can support their growing understanding.

Moving forward by encouraging predictions

A second type of reflection problem might involve students in generalizing and applying their initial understanding from concrete examples in the making of predictions. Figure 2.7 presents a sample problem that invites students to take this next step. However, before examining this problem, think about the question in Reflect 2.2.

Reflect 2.2

Imagine that you have given your students a variety of reflection tasks, some with the line of reflection oriented vertically, some with it oriented horizontally, and some with it oriented obliquely on the page.

How do you think you might help your students gain a better understanding of the similarities and differences between working with and locating reflections over a line that is vertical or horizontal and doing the same thing with reflections over a line that is oblique?

Locating reflections over a line presented in an oblique orientation is the challenge for students in the problem in figure 2.7. This problem calls on them to use visual and spatial reasoning to make predictions about the location of several reflections that are determined by a line that lies diagonally on the page.

Examine line *h* and the three objects labeled A, B and C.

- Predict where object A will end up if it is reflected over line *h*. Mark your prediction with an *a*. Explain how you located your prediction.

- Predict where object B will end up if it is reflected over line *h*. Mark your prediction with a *b*. Explain how you located your prediction.

- Predict where object C will end up if reflected over line *h*. Mark your prediction with a *c*. Explain how you located your prediction.

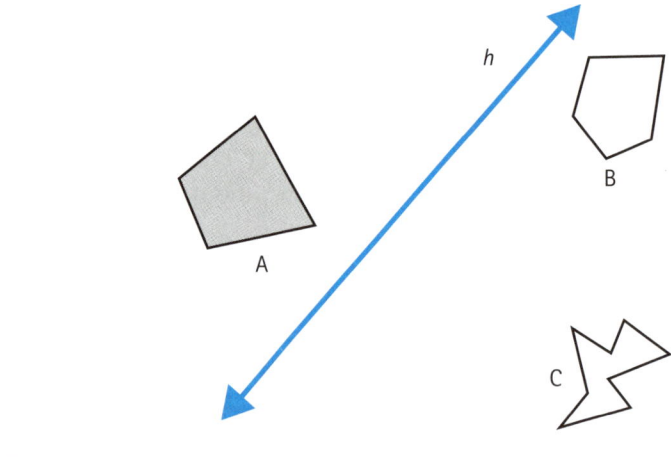

Fig. 2.7. A second type of reflection problem: Making predictions about reflections of objects A, B, and C across line *h*. Adapted from Slovin and colleagues (2003, p. 13).

Predicting the outcome of a problem is a good way for students to build an understanding of the elements involved in the mathematics. Making and then testing a prediction can help them develop strategies for solving a problem and assessing the results, a process that is related to Mathematical Practice 3 in CCSSM: "Construct viable arguments and critique the reasoning of others" (NGA Center and CCSSO

2010, p. 6). The sample problem in figure 2.7 is intended to prompt students' thinking about how the given orientation of the line of reflection affects their location of the reflected image. Many students' only experience with reflected images is with lines of reflection that they encounter in either a vertical or a horizontal orientation. Lines that are neither horizontal nor vertical provide more challenge as students try to envision the distance between the object and the line and the angle of the object in relation to the line.

The problem in figure 2.7 expects students to make their predictions without actually performing the reflections. Because it does not ask them to draw or sketch the images of the objects but only to make predictions, a discussion of the students' thinking can focus on the critical issues of distance and direction without the distraction of issues of accuracy or inaccuracy in the students' drawings. In this problem, one of the pre-images lies on one side of the reflection line, and the other two pre-images lie on the other side of the line. The entire plane on one side of the line of reflection is paired with points on the other side of the line, and vice versa. The locations of the pre-images relative to the line of reflection suggest a number of questions—for example, "Why will the images of the three figures be at different distances—closer to or farther away—from the line of reflection?" Students may account for these differences by explaining that when you reflect the figure, you reflect the space around it, too.

This problem also provides an opportunity to discuss the role of the line of reflection. Every reflection (which includes the plane containing the pre-image and the image) has a unique line of reflection. In the problem in figure 2.7, the line of reflection is oriented diagonally. This presentation is intended to challenge students, potentially eliciting an error that students commonly make in locating images when they encounter a line of reflection that is not parallel to an edge of the paper. In such circumstances, they often ignore the angle of the line of reflection and instead move horizontally or vertically on the paper. Figure 2.8 shows an example of such an error by a student who was attempting to locate the reflection of object B. The student incorrectly believed that the image of object B should be located in the area indicated by object B′. (It is worth telling students that, by convention, the image of an object is often given the same name as the pre-image, with the addition of a prime symbol ['].) During a class discussion, you can use errors such as these to have students debate and test one another's predictions.

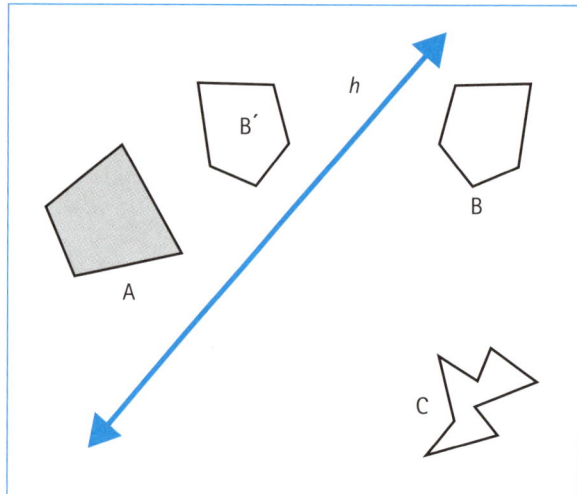

Fig. 2.8. Incorrect representation of B′, the reflected image of object B, across line *h*

After the students have had a chance to discuss their predictions, they should carry out the reflections. Their predictions, the actual locations, and the resulting discussion will help students focus on some of the properties involved in the transformation.

If you find that some of your students are having great difficulty making predictions and testing them, you might suggest that they turn their paper so that they are viewing the line of reflection oriented vertically on the paper. Have the students sketch their predictions and then turn the paper back to its original alignment. This scaffolding strategy should help students gain a better sense of the relationship between the line of reflection and the objects. Students are unlikely to make the choice illustrated in figure 2.8, for example, if they first rotate the line of reflection and inspect it in a vertical orientation, as illustrated in figure 2.9. They can more readily see that the object sketched as B′ is not the image of object B reflected over line *h*.

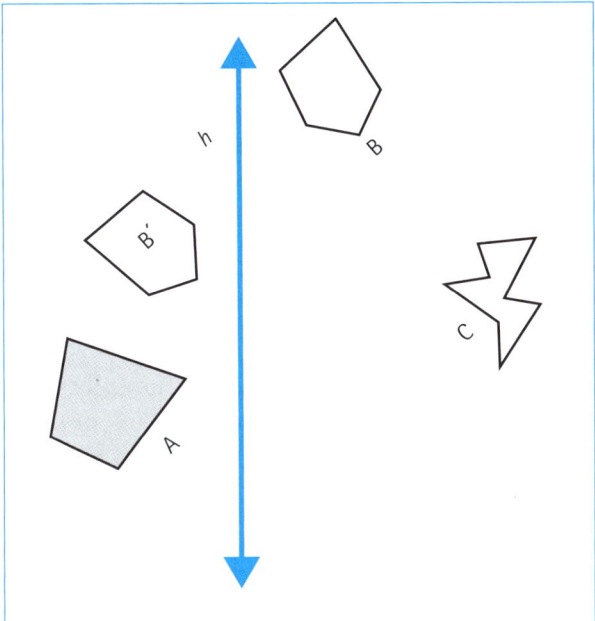

Fig. 2.9. The reflection problem from figure 2.7, with the error shown in figure 2.8, turned to present line *h* in a vertical orientation

To extend your students' experience with the relationship between the line of reflection and the reflected object or objects, you might draw an object or simple shape and ask them to draw a line of reflection anywhere they wish and complete the reflection. Next, they could work with the same pre-image, but this time draw a line of reflection that will result in a reflected image that is farther away from the line than the first reflected image. As a final step, they could draw a third reflection line that will result in a reflected image that is closer to the line of reflection than the first image was.

This work should be followed by a whole-class discussion in which students present their drawings and attend closely to the way that the distances between the lines of reflection and the images vary, depending on the position of the line of reflection relative to the pre-image. Although there will be variation in how far from or close to the lines of reflection different reflected images lie, the discussion should lead the students to make generalizations about the relationship between them. Focusing attention on this relationship can also draw attention to the reflection of the plane, not just the objects.

Students can do the activity with tracing paper and pencil; or if you have computers available, they can use GeoGebra, a public domain software program (https://www.geogebra.org/). As students become more familiar with the properties of reflections, you can include problems that allow them to continue to develop their understanding of the effect of relative positions of the line of reflection and the pre-image on the location of the final image.

Reflecting several objects over the same line to reinforce the idea of reflecting the plane

After students have worked with everyday reflection experiences and have made predictions about the locations of images when the line of reflection is in different orientations, they are ready for a task that offers a new challenge while cementing the idea that the entire plane—not just one or more objects of local interest—is transformed under a reflection.

The third task in the sequence engages students in reflecting several objects over the same line of reflection. The objects themselves are more complicated than those in the second task, and they are strategically positioned to deepen students' understanding of the idea that the entire plane is flipped over the line in a reflection. Students' sense of this idea is heightened when the pre-images lie on opposite sides of the line of reflection. Tasks like the third problem suggested for the sequence build on discoveries that students have made about the role of the line of reflection in their previous work of determining the location and position of a reflected image. However, a problem like that in figure 2.10 poses an additional challenge. Examine the task while focusing on the questions in Reflect 2.3.

Reflect 2.3

The task in figure 2.10 asks students to reflect three objects, Q, R, and H, over line *m*.

What unique challenges do you think each object would present?

Which object do you think would be most challenging for students to reflect?

How do the locations of the objects support a deeper understanding of reflections?

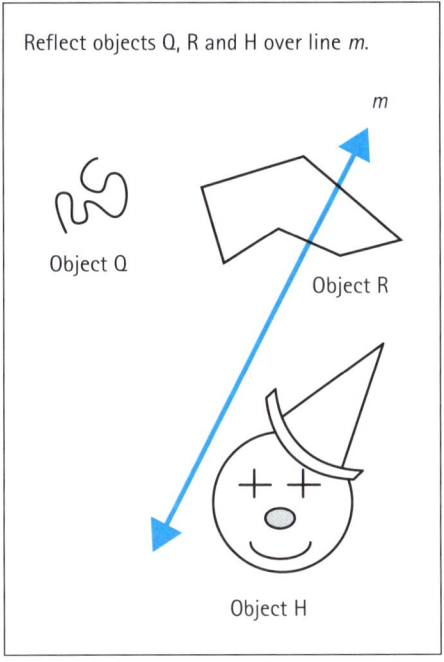

Fig. 2.10. A third-level reflection problem, with objects lying on both sides of the line of reflection. From Slovin and colleagues (2003, p. 39).

In the problem in figure 2.10, the line of reflection runs through the pre-image of object R. Before giving students this problem, you might want to review the main ideas of the problem in figure 2.7 or another problem where students draw different lines of reflection. Ask students to make a conjecture about how far away a pre-image can be from the line of reflection. Also ask them how close the pre-image can be to the line of reflection. These types of prediction tasks support students in developing generalizations about the relationship between the line of reflection and the locations of the pre-image and the image.

Students may have some difficulty completely reflecting object R. Many may say that the reflection cannot be accomplished because the line of reflection runs through the object. Although students may understand that the distance from the image to the line will be equal to the distance of the pre-image to the line, they may not be able to draw on their previous experience to reflect object R. Experience with a mirror, or another reflective device other than a Mira, may not be helpful. Some students may ignore the part of the object on one side or the other side of the line of reflection, depending on how they create the reflection. Others may realize that they need to show a reflected image of the entire object, but their mental picture of the image may not be clear enough for them to make the image, or they may be unable to figure out how to accomplish the reflection.

If students have access to a Mira, you can suggest that they look at the image from both sides of the tool to complete the reflection. If, in addition to a Mira, they have tracing paper, you might suggest that they fold a sheet and place the fold along the line of reflection, use the Mira, and again trace the image from both sides of the Mira. In a subsequent discussion about this problem, it would be useful to have students choose and label corresponding points on each pre-image and reflected image and connect them. You should ask students what they notice about the line that connects each pair of corresponding points (it is perpendicular to the line of reflection and is bisected by it) and what they notice about how all the lines connecting these pairs are related (they are parallel to each other). Consider these ideas as you respond to Reflect 2.3.

Reflect 2.4

Reexamine the task in figure 2.10.

Does any part of the image of object H lie on line *m*? Any part of object Q? Any part of object R?

What generalization can you make from these objects and their reflections?

This suggested third problem in the sequence presents an opportunity to introduce the idea of the fixed points of a transformation—points that are paired with themselves. A great deal of information can be gleaned about a transformation by examining its fixed points. Because every point on the line of reflection is a fixed point, we know that the image of the object presented as R in figure 2.10 must contain the points of intersection of object R and line *m*—points *A* and *B* in figure 2.11.

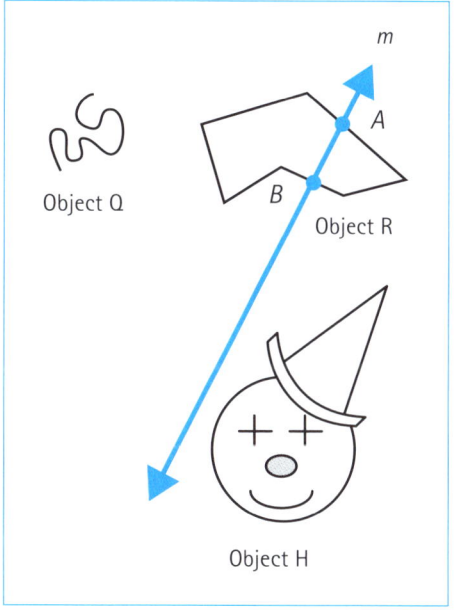

Fig. 2.11. The image of object R must also contain points *A* and *B*.

Assessing understanding of reflections

After students have solved and discussed problems like the three examples presented previously, a last reflection task can serve a twofold purpose. First, it allows students to summarize and reinforce their thinking that reflections involve the entire plane by describing specific ways in which these facts can be tested. Second, it gives teachers the opportunity to assess their students' understanding. The students' statements, as well as the types of characteristics on which they focus, provide information about the depth of the students' thinking. Do their statements focus on superficial features, or do they identify and explore relationships among the pre-image, image, and line of reflection? Consider the task in figure 2.12, for example.

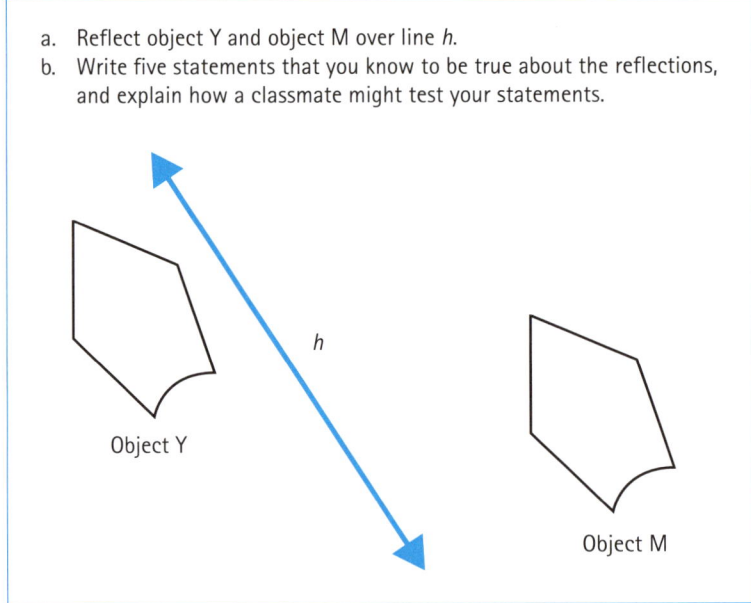

a. Reflect object Y and object M over line *h*.
b. Write five statements that you know to be true about the reflections, and explain how a classmate might test your statements.

Fig. 2.12. A final reflection problem: Assessing students' understanding of reflections. From Slovin and colleagues (2003, p. 40).

The reflection problem in figure 2.12 can serve as a useful assessment of students' understanding. It is likely to be accessible to most middle-grades students who have worked with reflections. Although the reflections that the students are asked to complete are not particularly challenging in themselves, the methods that students use to complete them, accompanied by their responses to part (b), can reveal a great deal about their conceptual understanding of reflections. For example, students who complete part (a) by reflecting each object separately may not yet view the two objects as part of the entire plane. Responses to part (b), in which students list what they know to be true about the situation and justify that knowledge, will show the depth and complexity of their understanding. The problem invites students' attention to certain details and general properties. Most notably, although objects M and Y are congruent and angled identically with respect to the line of reflection, their respective images highlight the distances between the objects and the line of reflection are reflected along with the objects. Students may also talk about corresponding vertices and line segments.

A cautionary note about transformations presented only in the coordinate plane

Restricting the types of problems that students encounter can add to students' misconceptions about transformations. Many middle-grades textbooks present transformations only in the context of the coordinate plane. This approach emphasizes a comparison of the properties of the pre-image and the image and describes the change in location by numbers—that is, by coordinate pairs. Students who have this limited experience tend to develop a static conceptualization of transformations, noting the "before" and "after" without developing an appreciation of the dynamic motion itself.

For example, in solving a problem in which an object whose location is identified by its coordinates on a graph is reflected over the y-axis, students can "correctly" solve the problem without attending to the reflection of the entire plane. They can relocate the reflected image vertex by vertex, merely counting and matching how many units from the axis each point is without giving any consideration to the motion itself or that the transformation affects anything besides the obvious figure.

Moreover, reflecting the coordinate plane across a line that is not horizontal or vertical presents additional difficulties. As figure 2.13 suggests, it is not a trivial problem to find the reflection of $\triangle ABC$ across line s.

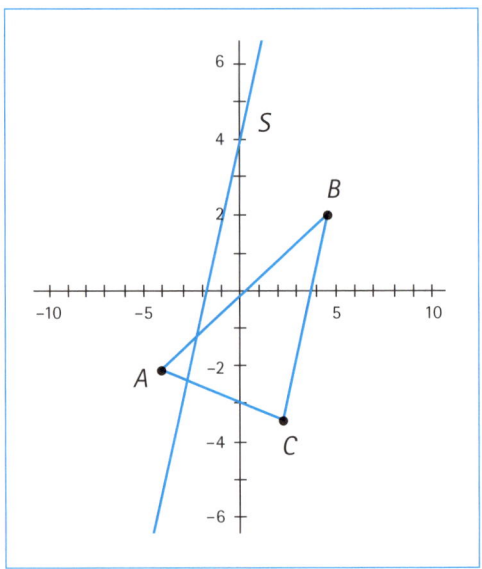

Fig. 2.13. Finding the image of △ABC in a reflection across line s.

Working with rotations, translations, and dilations

The previous discussion aimed to offer a fairly thorough treatment of many issues related to the teaching and learning of reflections in the middle grades. Effective teaching of rotations, translations, and dilations could be covered with the same degree of detail. However, the following discussion illustrates only a few of the topics associated with these transformations, while highlighting ideas that are commonly challenging for students in grades 6–8. You are encouraged to approach the teaching of these transformations in a manner similar to that previously illustrated for reflections.

Rotations: Turning the entire plane

Wesslén and Fernandez (2005) found that some middle-grades students—particularly lower-achieving students in year 8—

> were not confident with rotating shapes where the center of rotation was not on the edge or middle of the shape to be rotated. Again, this shows students do not think of rotation as a rotation of the plane, and as a way of moving a shape to a new location, but merely as a way of rotating a shape but keeping it in the same place. (p. 29)

Figure 2.14 presents a rotation problem accompanied by a student's solution. Consider the problem and the solution, and then explore the questions in Reflect 2.5.

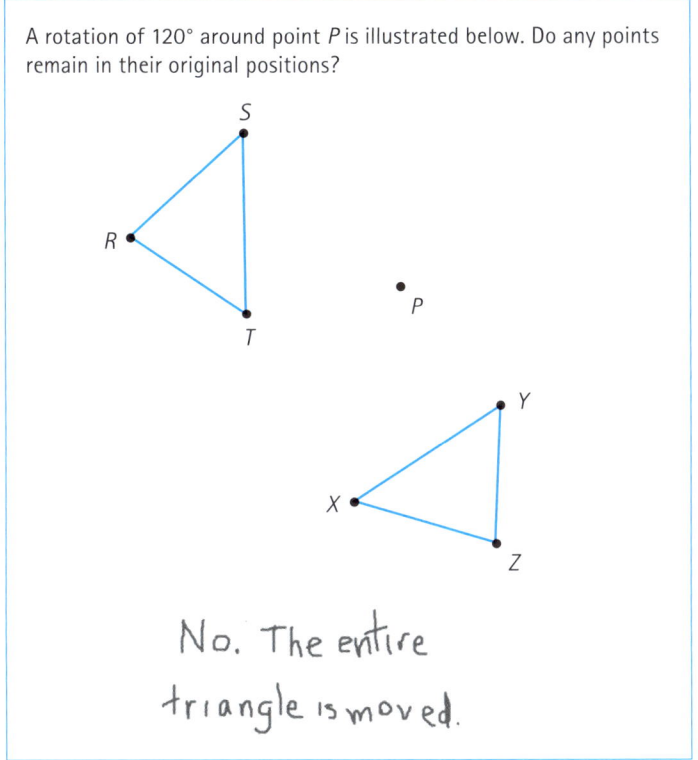

A rotation of 120° around point *P* is illustrated below. Do any points remain in their original positions?

No. The entire triangle is moved.

Fig. 2.14. A rotation problem and a student's response

Reflect 2.5

What would you say to one of your students who gave the answer shown in figure 2.14?

How would you help this student develop a better understanding of the properties of a rotation?

How do you think your students would respond if *P* were on △*RST*? In the interior of △*RST*?

Do you think presenting both △*RST* and △*XYZ* concurrently makes the problem easier or more difficult?

Students can use tracing paper to give them a better understanding of why the point *P* is not moved by the rotation. They could trace △*RST* and point *P* and then use a pencil to hold *P* fixed while turning the paper 120 degrees (you might need to remind your students that a positive angle measure describes a counterclockwise rotation). You might also want to have students add other figures (some containing *P*) to the original page and then use tracing paper in the same way to help them realize that the rotation affects all the points in the plane—not only the points that make up △*RST*.

Typical rotation problems give students a pre-image object, a center of rotation, and an angle of rotation and ask them to find the image. However, students in grades 6–8 should also have experience with a reverse problem—one that gives them a pre-image object and its image and asks them about the rotation. Such a problem is shown in figure 2.15. Examine the problem, and then consider the hypothetical situation and the associated question in Reflect 2.6.

Hexagon *A´B´C´D´E´F´* is the image of hexagon *ABCDEF* under a rotation of the plane. Which of the points *V, W, X, Y,* or *Z* is the center of this rotation? What is the angle of rotation? Explain your reasoning.

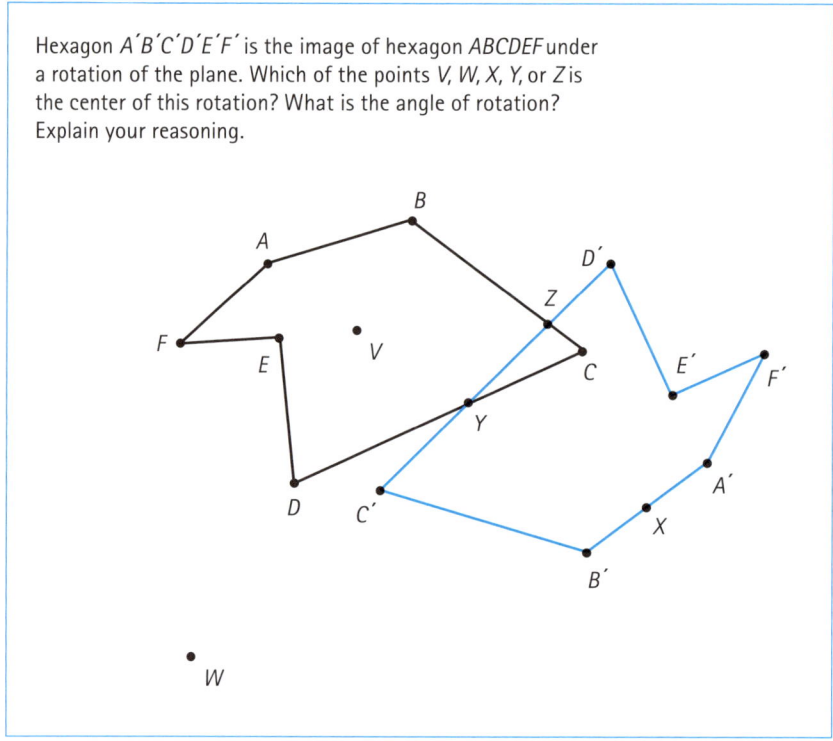

Fig. 2.15. Finding the center and angle of a rotation

Reflect 2.6

Referring to the rotation problem in figure 2.15, suppose that one of your students says that the center of rotation is either point *V* or point *W* because "That's the way it usually looks in the textbook," and another student disagrees and asserts that the center must be either point *Y* or point *Z*.

How would you respond to these two students?

A conceptual understanding of rotations would lead a student to conclude that either *Y* or *Z* is the center of the rotation, since those are the only two points that lie on both the image and pre-image hexagons. This would imply that these two points are the only ones that could possibly remained fixed by the rotation and therefore could be its center. The student could determine which of these is actually the center of the rotation by examining some angles. If *Z* is the center, then the angles *AZA′*, *BZB′*, *CZC′*, and so on must all be congruent. If *Y* is the center, then the angles *AYA′*, *BYB′*, *CYC′*, and so on must all be congruent. A look back at the diagram in the problem would make it apparent that *Y* is the center of the rotation, with an angle of rotation congruent to $\angle AYA'$, or 200 degrees.

Translations: Understanding congruent translation vectors

A difficulty that students often have in middle-grades work with translations is understanding that the translation vector can appear anywhere in the plane, since two vectors are congruent if they have the same direction and length. A problem such as that in figure 2.16 can illustrate and highlight this idea for students.

The plane is translated by using vector *a* (shown as an arrow with a given direction and length). As a result of the translation, the point *P´* is the image of the point *P.* Vectors *a* and *b* have the same direction and length. What would be the image of the point *P* if vector *b* were used to translate the plane instead of vector *a*?

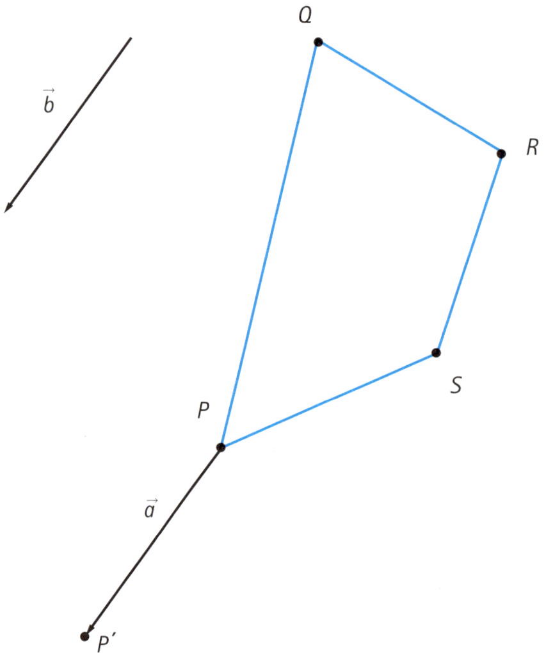

Fig. 2.16. A translation of the plane defined by two congruent vectors, *a* and *b*

This misunderstanding about the translation vector appears to be fairly common and occurs across the spectrum of students. Wesslén and Fernandez (2005) expressed surprise about how pervasive they found it to be:

> Even some more able students did not realise that translating a shape with reference to one point or another gives the same result. This indicates a fundamental misconception as many students did not seem to understand that a translation moves the whole plane—and whatever happens to be on it—the same distance and direction. (p. 28)

Thus, many students may not understand that the image of *P* is the same, regardless of whether vector *a* or *b* is used. A common misconception among students is that if vector *b* is the translation vector, they must first move quadrilateral *PQRS* so that *P* will coincide with the tail of vector *b* before they can find the image *P´* of *P*. A sample student response is shown in figure 2.17.

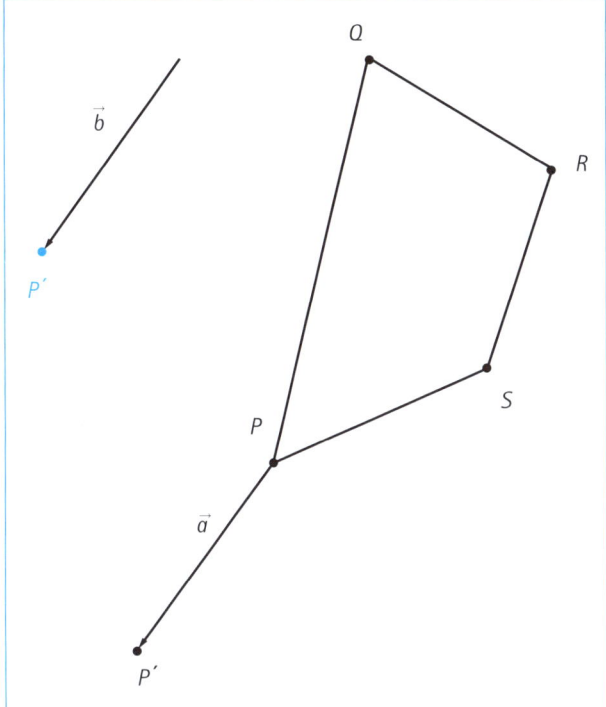

Fig. 2.17. A student's misconception of the location of the image of *P* under the translation defined by vector *b*

Dilations: Scaling to enlarge, reduce, or preserve size

A dilation is a transformation that preserves similarity. That is, a pre-image object and its image have the same shape. However, congruence may or may not be preserved, and a dilation is not, therefore, an isometry, which, as noted earlier, always preserves size as well as shape. In a dilation, the image object may be larger than, the same size as, or smaller than the pre-image object, depending on whether the scaling factor is a positive number that is greater than, equal to, or less than 1, respectively. The differences between dilations and the isometries thus may be misunderstood. Even the name *dilation* itself may cause students to believe that the image of an object that results in a dilation transformation is always larger in size than the pre-image object, since the words *dilate* and *dilation* are often used to mean "grow or make larger" and "enlargement."

Dealing with scaling factors and the arithmetic of ratios and proportions can be another source of confusion for middle-grades students. Consider the sample problem in figure 2.18.

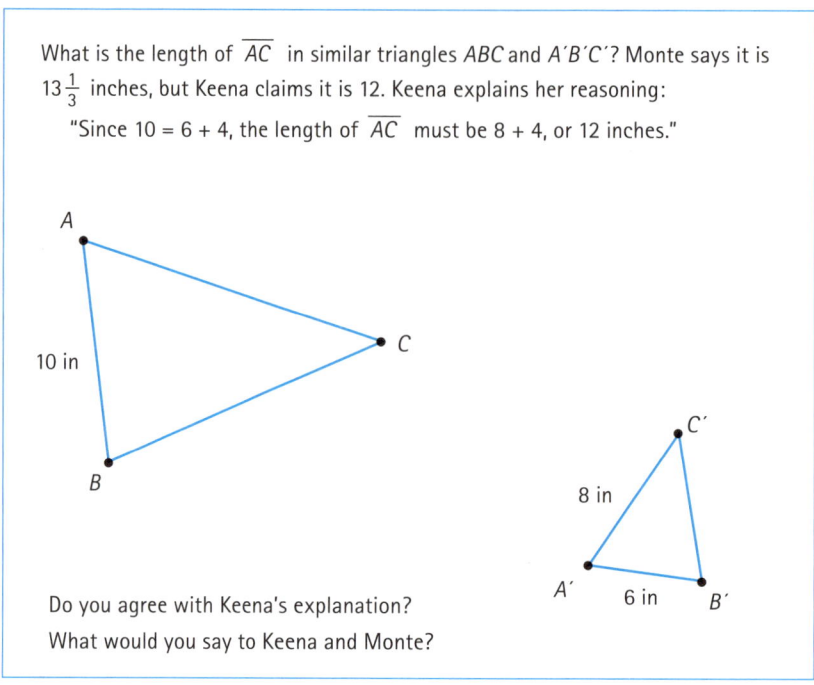

What is the length of $\overline{AC}$ in similar triangles ABC and $A'B'C'$? Monte says it is $13\frac{1}{3}$ inches, but Keena claims it is 12. Keena explains her reasoning:

"Since 10 = 6 + 4, the length of $\overline{AC}$ must be 8 + 4, or 12 inches."

10 in

Do you agree with Keena's explanation?

What would you say to Keena and Monte?

8 in

6 in

Fig. 2.18. Finding the length of $\overline{AC}$ in similar triangles ABC and $A'B'C'$

Keena's explanation illustrates middle school students' frequent use of additive reasoning when they are attempting to work with ratios and proportions. In this case, Keena thought

$$\frac{6}{8} = \frac{10}{12},$$

basing her idea of the equivalence of these two ratios on the fact that both the numerators and denominators differ by 4. A discussion of the problem should reinforce the idea that scaling factors have a multiplicative structure by analyzing Monte's response:

$$AC = 13\frac{1}{3} \text{ inches, since } \frac{6}{8} = \frac{10}{13\frac{1}{3}}.$$

Figure 2.19 shows $A'B'C'D'E'$ as the image of pentagon $ABCDE$ under a dilation centered at P. Four lengths are labeled in the diagram. Giving students a dilation labeled in this way can be useful in bringing to the surface underlying confusion and misunderstanding that students often harbor about the ratios involved in a dilation. The figure also presents a discussion between two students, Hannah and

Gina, who are trying to understand the lengths given for two corresponding sides of the pre-image and image pentagons. Consider the students' discussion, and respond to the questions in Reflect 2.7.

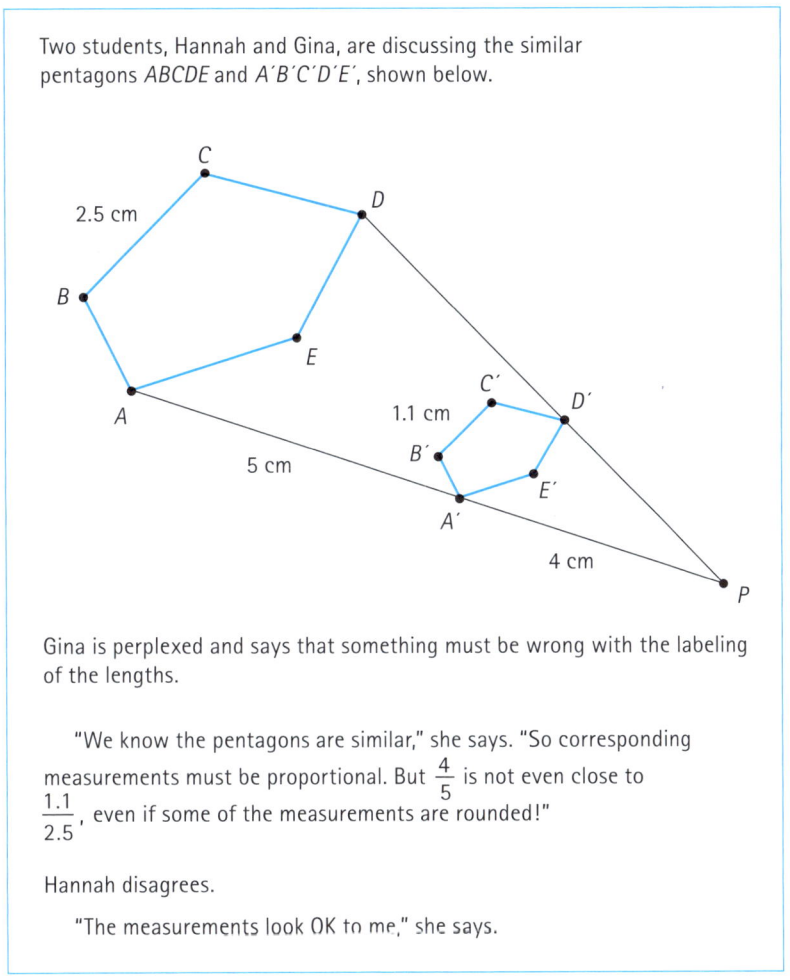

Two students, Hannah and Gina, are discussing the similar pentagons *ABCDE* and *A′B′C′D′E′*, shown below.

Gina is perplexed and says that something must be wrong with the labeling of the lengths.

"We know the pentagons are similar," she says. "So corresponding measurements must be proportional. But $\frac{4}{5}$ is not even close to $\frac{1.1}{2.5}$, even if some of the measurements are rounded!"

Hannah disagrees.

"The measurements look OK to me," she says.

Fig. 2.19. *A′B′C′D′E′* is the image of *ABCDE* under a dilation with center at *P*

Reflect 2.7

Reexamine figure 2.19, and analyze Gina and Hannah's disagreement about whether the lengths labeled in the dilation can be accurate.

Who is correct, Gina or Hannah? Why?

What would you say to Gina and Hannah?

Gina's thinking illustrates a common misconception that students have about dilations and similar figures. Students may understand that because the two pentagons are similar to each other, corresponding sides must be proportional. Further, they may recognize that the equal ratios in the proportion represent the scaling factor; hence, for this dilation, the scaling factor is

$$\frac{PA'}{PA}, \quad \text{not} \quad \frac{PA'}{A'A}.$$

However, in looking at the diagram, they may not understand that the scaling factor for the dilation is defined to be

$$\frac{PA'}{PA} = \frac{4}{9} = 0.44,$$

as Gina surmised. Because

$$\frac{1.1}{2.5} = 0.44,$$

Hannah is correct that the measurements given in the labels appear to be fine.

Summarizing Pedagogical Content Knowledge to Support Big Idea 3 and Essential Understanding 3*b*

Teaching the mathematical ideas in this chapter requires specialized knowledge related to the four interrelated components presented in the Introduction: learners, curriculum, instructional strategies, and assessment. The four sections that follow summarize some examples of these specialized knowledge bases in relation to Big Idea 3 and Essential Understanding 3*b*.

Knowledge of learners

Many students come to the middle grades with an informal, intuitive knowledge of transformations. They have been flipping, turning, sliding, and changing the sizes of toys, blocks, shapes, and other objects since early childhood. Transformations are simply models for conducting these everyday activities. As the four problems presented in the section on reflections illustrate, students can master the concept of transformations by exposure to problem situations that first draw on their intuitions from previous, everyday experiences and later add opportunities to make and test conjectures before they are asked to provide informal justifications for their results.

As they progress along the learning trajectory, students benefit from using geometric tools such as concrete models, tracing paper, templates, Miras, protractors, and dynamic geometry software to explore and solve problems. As they advance, students should also develop a conceptual understanding of transformations that gives them the ability to work abstractly, without needing pictorial representations of the plane, the use of coordinates, or the making of numerical measurements.

Knowledge of curriculum

Selecting appropriate instructional tasks is critical to helping students develop a solid understanding of transformations. Teachers must take care to ensure that they offer tasks that present important features of transformations in a variety of ways. For example, as discussed, reflection tasks should include lines of reflection that are drawn horizontally, vertically, and obliquely. Furthermore, the lines of reflection should sometimes intersect the given objects. Centers of rotation and dilation should sometimes be on or in the interior of the given objects, as well as external to them. Translation tasks should provide students with opportunities to see translation vectors of varying lengths and directions. Also, dilation tasks should ensure that students see examples in which the scaling factor is between 0 and 1, as well as those in which it is greater than 1.

A key idea for teachers to bear in mind when considering transformation tasks for middle-grades students is that tasks must be structured wisely to allow students to abstract the relevant characteristics of transformations while keeping superficial aspects in the background. In addition, it is important to remember that only through a wide variety of experiences can students develop a conceptual understanding of transformations that will allow them to move from a concrete to a visual and finally to an abstract knowledge of reflections, rotations, translations, and dilations.

Knowledge of instructional strategies

As is the case for much of mathematics, the most effective instruction in transformations is instruction that builds on the knowledge that students bring to the classroom. This type of instruction allows students to explore transformations on their own, naturally and informally, accompanied by the teacher's light touch. Students should spend time talking with one another, making and testing conjectures, and justifying their results. Incorporating dynamic geometry software, such as the Geometer's Sketchpad or GeoGebra, into your classroom practice is an important strategy that allows students to explore transformations more efficiently and make and test their conjectures.

Knowledge of assessment

Rose, Minton, and Arline (2007) give some general tips for using formative assessment probes effectively to uncover students' thinking. These tips include providing a problem that offers opportunities for (1) bringing students' understandings and misunderstandings to the surface, (2) probing and challenging students' understanding of a particular concept, and (3) examining students' approaches and thought processes. Burrill (2005) gives specific examples of items for assessing transformations. What is most important is assessment that focuses on students' thinking, explanations, strategies, and reasoning instead of strictly on the problems' solutions. Assessment that focuses on thinking can lead students to a better understanding of the underlying structure of transformations and their properties and will serve the students well in subsequent mathematics courses.

Conclusion

Chapter 2 has discussed content and pedagogy associated with the teaching and learning of transformations. Especially important is a carefully sequenced series of problems that take students from an intuitive understanding to making and testing conjectures, as illustrated by the series of sample problems detailed for reflections. The chapter has presented examples of student work and reflection questions to highlight important ideas for putting Big Idea 3 and Essential Understanding 3*b* into practice. Transformations are an important part of state and national mathematics standards and postsecondary school mathematics courses. Middle-grades students who develop a solid conceptual understanding of transformations will have established a foundation on which they can build in the future.

Chapter 3
Looking Back and Looking Ahead with Geometry

This chapter looks in opposite directions to highlight ways that the essential understandings discussed in chapters 1 and 2 align with the understanding of geometry that students develop before they arrive in grades 6–8 and after they move on to grades 9–12 and to postsecondary work in mathematics. The chapter begins with a discussion of understandings that students are expected to build in grades 3–5. When you encounter students in grades 6–8 who have gaps in their knowledge, you may need to assess their understanding of the ideas that this first section highlights. The second section discusses ways in which the essential understandings discussed in chapters 1 and 2 connect with and support the mathematics that students encounter beyond grade 8. Although these more advanced topics are too numerous to discuss exhaustively, the discussion in this section demonstrates how important it is for students in grades 6–8 to develop a deep understanding of the essential concepts that serve as a foundation for subsequent study of geometry in high school and beyond.

Building Foundations: Geometry in Grades 3–5

Lindquist and Clements (2001) encourage the inclusion of geometry in elementary grades for developing students' abilities to justify conclusions. Helping students investigate, make conjectures, and develop logical arguments as they move through the grades supports their development of spatial sense. In fact, Clements and Sarama (2011) argue that including strong geometry experiences in the elementary grades supports mathematics achievement in all areas.

Geometry experiences in grades 3–5 include a large component focused on geometric measurement as recommended in the Common Core State Standards for Mathematics (CCSSM; National Governors Association Center for Best Practices and Council of Chief State School Officers [NGA Center and CCSSO] 2010). In grade 3, finding the perimeters of polygons goes beyond counting the units on the boundary of a shape. Third graders are expected to be able to apply perimeter concepts and skills as noted in the following (NGA Center and CCSSO 2010, p. 25):

> Solve real world and mathematical problems involving perimeters of polygons, including finding the perimeter given the side lengths, finding an unknown side length, and exhibiting rectangles with the same perimeter and different areas or with the same area and different perimeters. (3.MD.8)

Some problems may involve incomplete drawings of a rectangle and its side measures, motivating students to apply their understandings of the characteristics of rectangles and perimeter. For example, a problem might be presented as in figure 3.1.

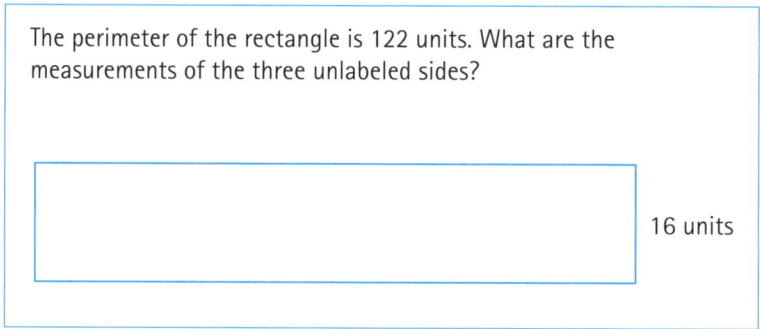

The perimeter of the rectangle is 122 units. What are the measurements of the three unlabeled sides?

16 units

Fig. 3.1. A problem calling for the unknown side lengths of a rectangle

To solve this problem, students have to recognize that opposite sides of a rectangle are the same length and coordinate this fact with the process of finding perimeter. The way in which this problem is presented forces students to move from simply adding the lengths of the sides to a higher level of thinking.

At this stage, students can also begin to tackle problems that are much more open-ended, such as this reversibility task:

> Find three rectangles whose perimeter is 24 inches.

There are multiple solutions to this problem, including rectangles that measure 1 inch by 11 inches, 3 inches by 9 inches, and 2 inches by 10 inches. Although the rectangles will be different, their perimeters will be the same. The pairing of visualization of these rectangles with the numerical aspects of finding perimeter continues to increase the complexity of problems that students solve at this grade level.

Although third graders' work with perimeter is a continuation of their work with geometric measurement in earlier grades, area is a new topic for these students and becomes a focus of their work, as indicated by the numerous standards for grade 3 related to area (NGA Center and CCSSO 2010, p. 25):

Geometric measurement: understand concepts of area and relate area to multiplication and to addition. (3.MD.5–7)

5. Recognize area as an attribute of plane figures and understand concepts of area measurement.

    a. A square with side length 1 unit, called "a unit square," is said to have "one square unit" of area, and can be used to measure area.

    b. A plane figure which can be covered without gaps or overlaps by n unit squares is said to have an area of n square units.

6. Measure areas by counting unit squares (square cm, square m, square in, square ft, and improvised units).

7. Relate area to the operations of multiplication and addition.

    a. Find the area of a rectangle with whole-number side lengths by tiling it, and show that the area is the same as would be found by multiplying the side lengths.

    b. Multiply side lengths to find areas of rectangles with whole-number side lengths in the context of solving real world and mathematical problems, and represent whole-number products as rectangular areas in mathematical reasoning.

    c. Use tiling to show in a concrete case that the area of a rectangle with whole-number side lengths a and $b + c$ is the sum of $a \times b$ and $a \times c$. Use area models to represent the distributive property in mathematical reasoning.

    d. Recognize area as additive. Find areas of rectilinear figures by decomposing them into non-overlapping rectangles and adding the areas of the non-overlapping parts, applying this technique to solve real world problems.

These standards connect spatial ideas with number topics as students develop concepts of multiplication through the presentation of area explorations. For example, given 24 square tiles, students may be asked to form as many rectangles as they can. At first, students may place the tiles one by one without noticing any pattern. They may then realize that the number of rows of tiles times the number of columns of tiles is the total number of tiles used to construct the rectangle. As they create more of the rectangles, they may generalize that the side lengths of the rectangles that are formed are the factors of 24. This generalization may then lead to the broader one that the area of a rectangle is equal to the product of the base and height of that shape.

In grade 4, students apply their concepts and skills related to area and perimeter in a variety of problems, as the Common Core State Standards recommend in the following standard (NGA Center and CCSSO 2010, p. 31):

> Apply the area and perimeter formulas for rectangles in real world and mathematical problems. (4.MD.3)

One mathematical problem that students might be given, for example, is the following reversibility task:

> Find the dimensions of three rectangles that have an area of 48 square units.

Another mathematical problem that they might explore is a flexibility task that asks them to find the perimeter and area of a rectangle in two different ways. Problems of these types, which call on students to apply their understanding of area and perimeter at the same time, promote a deeper knowledge of both concepts.

Real-world problems might include those that require students to rely on their spatial sense to decompose irregular shapes into more familiar ones, especially rectangles. For example, given the L shape in figure 3.2, students would need to determine the measurements of the sides that are not labeled as they decompose the shape.

Find the area of the shape below:

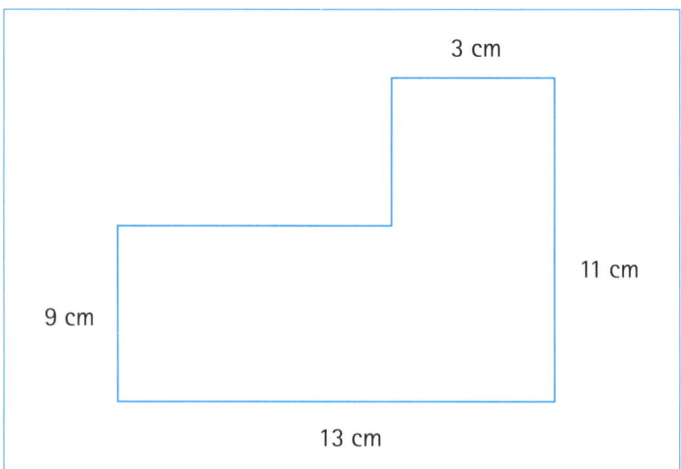

Fig. 3.2. A real-world problem calling for the area of an irregular shape

To solve this problem, students can decompose the shape in different ways. For example, some students might use one rectangle that measures 3 centimeters by 11 centimeters and another one that measures 9 centimeters by 10 centimeters. Other students might use a rectangle that measures 9 centimeters by 13 centimeters and another one that measures 2 centimeters by 3 centimeters. An underlying concept of this problem is that area is additive. In other words, the area of a composite shape is the sum of the areas of its parts. Furthermore, students have to visualize how to find the missing dimension of a rectangle in relation to the given measures. These two components of the problem raise the level of student thinking.

By grade 5, students use their understanding of perimeter and area related to two-dimensional shapes to begin explorations with volume. Up to this point, students have had to coordinate linear measures with area, but working with volume requires that they coordinate both linear and area measures in relation to the three-dimensional space that a shape occupies. Visualizing volume becomes more sophisticated activity, students' skill builds on their spatial sense from the previous two grades.

The Common Core State Standards for grade 5 that address geometric measurement focus specifically on the formalization of a formula for volume. The progress that students are expected to make from concept to skill is evident in the following standards (NGA Center and CCSSO 2010, p. 37):

Geometric measurement: understand concepts of volume and relate volume to multiplication and to addition. (5.MD.3–5)

3. Recognize volume as an attribute of solid figures and understand concepts of volume measurement.

    a. A cube with side length 1 unit, called a "unit cube," is said to have "one cubic unit" of volume, and can be used to measure volume.

    b. A solid figure which can be packed without gaps or overlaps using n unit cubes is said to have a volume of n cubic units.

4. Measure volumes by counting unit cubes, using cubic cm, cubic in, cubic ft, and improvised units.

5. Relate volume to the operations of multiplication and addition and solve real world and mathematical problems involving volume.

    a. Find the volume of a right rectangular prism with whole-number side lengths by packing it with unit cubes, and show that the volume is the same as would be found by multiplying the edge lengths, equivalently by multiplying the height by the area of the base. Represent threefold whole-number products as volumes, e.g., to represent the associative property of multiplication.

    b. Apply the formulas $V = l \times w \times h$ and $V = b \times h$ for rectangular prisms to find volumes of right rectangular prisms with whole-number edge lengths in the context of solving real world and mathematical problems.

    c. Recognize volume as additive. Find volumes of solid figures composed of two non-overlapping right rectangular prisms by adding the volumes of the non-overlapping parts, applying this technique to solve real world problems.

From their explorations, students may also begin to make generalizations related to the additive nature of perimeter, area, and volume as well as to other characteristics of these measures. For example, they may notice that perimeter is related to a 1-dimensional measure, area to a 2-dimensional measure, and volume to a 3-dimensional measure. Students may also see that a shape that can be measured with respect to area can also be measured with respect to perimeter or that a shape that can be measured with respect to its volume can also be measured in a different way, with respect to its surface area.

In much the same way that explorations help students develop concepts related to area, they also move students from counting cubes to find the volume of a rectangular prism to using a formula. While students move toward the formula, they begin to coordinate counting techniques and generalize patterns. Moving beyond counting cubes, they may notice, for example, that the number of cubes in the height times the number of cubes in the length times the number of cubes in the width is the same as the number obtained by counting the cubes one by one. Other generalizations may include noting that for a rectangular prism, replicating the rectangle at the base of the prism produces the prism. Thus, the area of the base times the height also leads to the volume.

As in their explorations of area, students investigate shapes that are irregular while they explore volume, using their spatial sense to decompose 3-dimensional shapes into rectangular prisms. Thus, they are able to make connections from 2-dimensional geometric concepts and skills to those involving 3-dimensional shapes.

By the end of grade 5, students should be relatively adept at composing and decomposing 2- and 3-dimensional shapes and applying formulas to find a geometric measurement. The ability to do this provides a foundation for future work in solving problems involving more complex shapes.

Extending Understanding: Geometry in Grades 9–12

As students begin to discover in grades 6–8, significant relationships exist among area, perimeter, volume, and transformations. Area, perimeter, and volume, treated as geometric measurements, have some similar characteristics, as shown in chapter 1. These characteristics are foundationally important, but students' sophistication increases while they move to grades 9–12, where the use of transformations to consider other relationships becomes more prominent.

The problems presented in chapter 2 foreshadow the important role that transformations play in geometry in grades 9–12 and beyond. Transformations feature prominently in the expectations set out for high school geometry in the Common Core State Standards for Mathematics. Transformations are also a foundational and unifying concept in higher mathematics. This section presents and discusses examples of ways in which transformations appear in the secondary school curriculum, with the goal of familiarizing middle-grades teachers with the geometry that their students are likely to encounter after they leave the middle school classroom.

Three transformations that students should have encountered when they enter high school are the isometries discussed in chapter 2—reflections, translations, and rotations. As observed previously, working with these transformations offers middle-grades students an intuitive, informal way of thinking about geometry as a natural part of their world. As middle-grades students discover, a reflection flips all the points on one side of a line to a corresponding location on the other side of the line (see fig. 3.3a). Triangle A is reflected across line *t* to give triangle B. Likewise, a translation slides all the points in the plane a specified distance in a specified direction. Figure 3.3b provides an example: Square C is translated to square D, as defined by the vector *PQ*. A rotation turns all the points in the plane around a given point, called the *center of rotation*, by a specified angle. Figure 3.3c shows quadrilateral E rotated 90 degrees counterclockwise around point *M* to give quadrilateral F. (As noted in chapter 2, a counterclockwise rotation is considered by convention to be motion in a positive direction.)

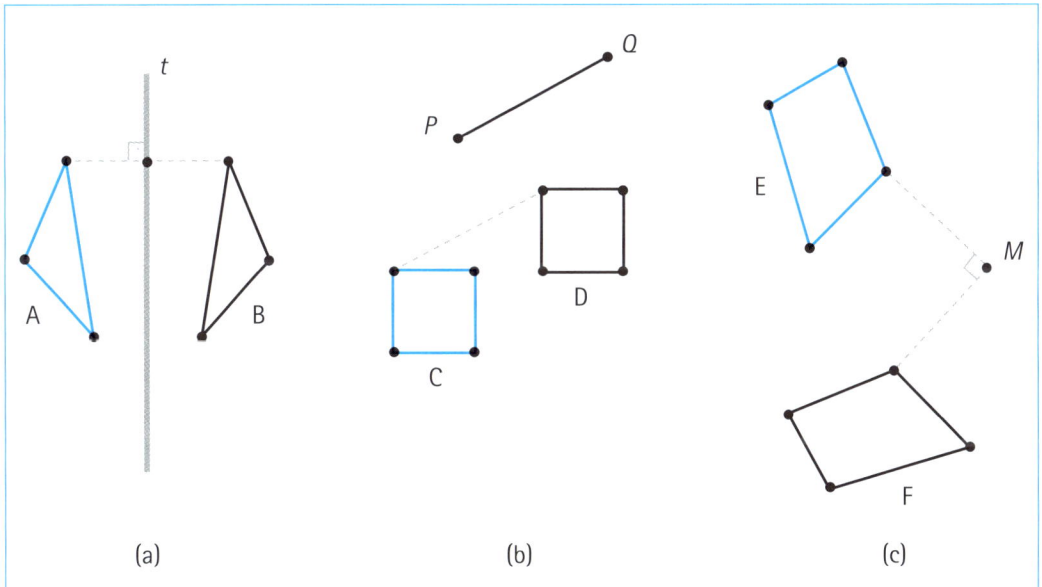

Fig. 3.3. Examples of isometries: (a) a reflection, (b) a translation, and (c) a rotation

Students who have rich experiences with isometries in grades 6–8 will reap the benefits in high school and beyond as they expand their work with transformations. The importance of understanding transformations is emphasized in the standards for high school geometry elaborated in CCSSM. High school students are expected to have understanding and skill enabling them to do the following (NGA Center and CCSSO 2010, p.76):

> Represent transformations in the plane using, e.g., transparencies and geometry software; describe transformations as functions that take points in the plane as inputs and give other points as outputs. Compare transformations that preserve distance and angle to those that do not (e.g., translation versus horizontal stretch). (G-CO.2)

> Given a geometric figure and a rotation, reflection, or translation, draw the transformed figure using, e.g., graph paper, tracing paper, or geometry software. Specify a sequence of transformations that will carry a given figure onto another. (G-CO.5)

> Use geometric descriptions of rigid motions to transform figures and to predict the effect of a given rigid motion on a given figure; given two figures, use the definition of congruence in terms of rigid motions to decide if they are congruent. (G-CO.6.)

These standards for high school geometry focus on the important topic of congruence (G-CO), and the fundamental idea for students to grasp is that two objects are congruent if one object can (perhaps figuratively) be picked up and placed on top of the other by reflecting (flipping), translating (sliding), or rotating (turning) it as necessary. The two objects have the same shape and size. Conceptualizing congruence in this way is much more general than the students' previous ways of thinking about it, since it applies to all objects and is perhaps more natural than the triangle congruence theorems students may have been asked to memorize previously (side-side-side, side-angle-side, angle-side-angle, and angle-angle-side).

Examples can illustrate this transformational approach to demonstrating congruence. In a case such as that presented in figure 3.4a, $\triangle ABC$ is congruent to $\triangle DEF$ since the first triangle can be made to coincide with the second triangle by rotating it 120 degrees around point C and then reflecting it across line m. However, in a case such as that in figure 3.4b, the circle centered at point O is not congruent to the circle centered at point P, since no sequence of reflections, translations, or rotations will result in one of the circles exactly matching up with the other. Although the two circles have the same shape, they have different sizes and therefore are not congruent.

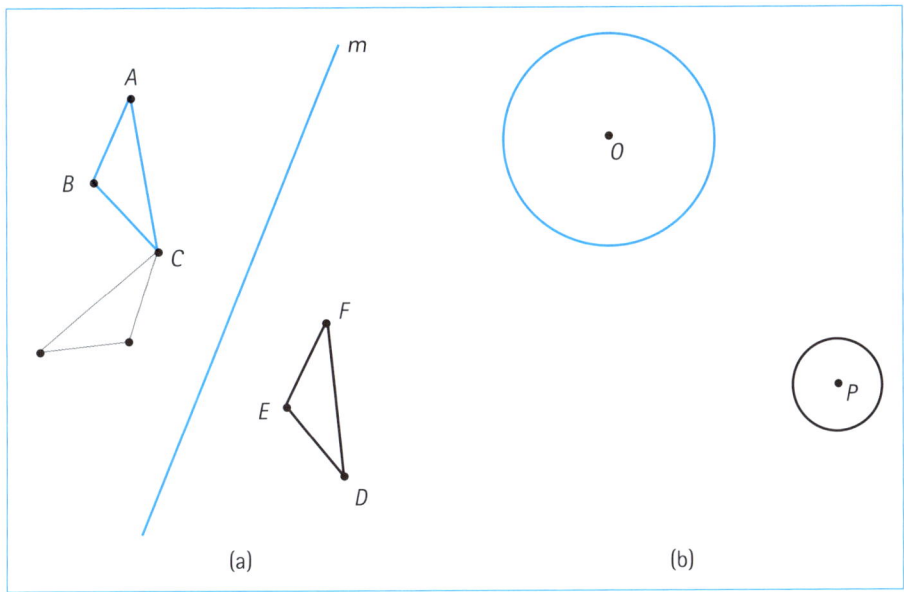

Fig. 3.4. Using transformations to determine congruence or a lack of congruence

As middle-grades students learn, a dilation is another, equally important transformation. A dilation can potentially change a figure's size but not its shape. Work
with dilations in grades 9–12 reinforces the idea that a dilation moves every point
in the plane to a new location in such a way that the ratio of its new distance from
a specified point to its original distance from that point is always a constant. As
students recall from their work in the middle grades, the specified point is called
the *center of the dilation*, and the constant is called the *scaling factor*. In figure 3.5,
the dilation centered at point *P* moves a point so that the ratio of its new distance
from *P* to its original distance from *P* is 1.85.

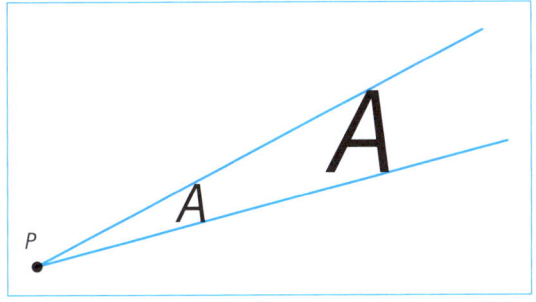

Fig. 3.5. A dilation centered at *P* with a scaling factor of 1.85

Figure 3.6 shows another object, $\triangle RST$, along with its image, $\triangle R'S'T'$, under the same dilation centered at point P. The addition of this object and its image reinforces the idea that every point in the plane, with the exception of P, is transformed in the dilation. Given the dilation centered at the point P with scaling factor 1.85, the small letter A is mapped to the large letter A, and $\triangle RST$ is mapped to $\triangle R'S'T'$ with

$$\frac{PQ'}{PQ} = \frac{PS'}{PS} = 1.85.$$

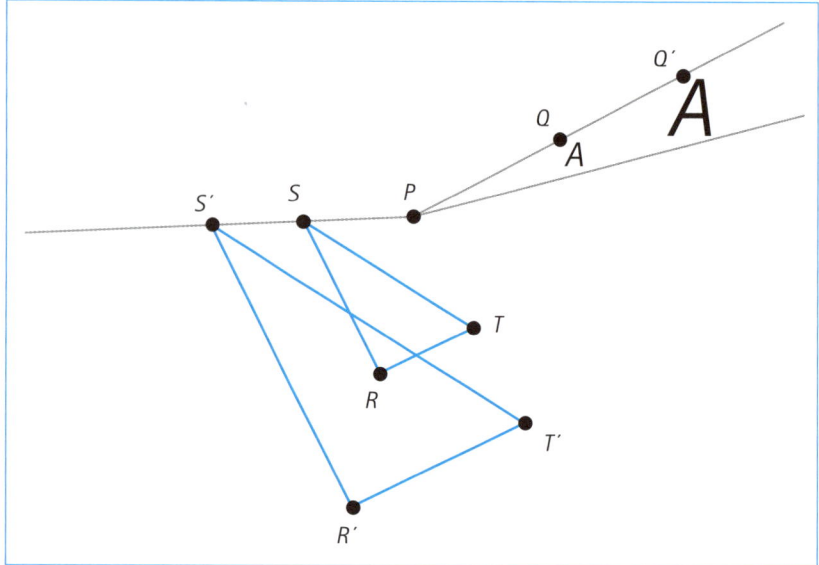

Fig. 3.6. A dilation centered at P, showing $\triangle R'S'T'$ with a scaling factor of 1.85

Whereas the isometries—reflections, translations, and rotations—can be associated with congruent figures, dilations can be associated with similar figures. Like isometries, dilations feature prominently in the CCSSM standards for high school geometry. In the case of dilations, however, the relevant standards focus on the important and interrelated topics of similarity, right triangles, and trigonometry (G-SRT; NGA Center and CCSSO 2010, p. 77). With respect to these topics, the following proficiency is central for high school students to develop and demonstrate:

> Given two figures, use the definition of similarity in terms of similarity transformations to decide if they are similar; explain using similarity transformations the meaning of similarity for triangles as the equality of all corresponding pairs of angles and the proportionality of all corresponding pairs of sides. (G-SRT.2.)

The abundance of rich mathematical ideas associated with isometries and dilations is remarkable. These transformations move lines to lines and parallel lines to parallel lines. They move angles to angles and perpendicular lines to perpendicular lines. Some transformations leave certain points and lines in the plane fixed, whereas others leave no point in the plane fixed. Transformations provide a context for exploring measurements of lengths and angles, proportional reasoning, area, perimeter, and, in high school geometry, functions, matrices, and algebra.

In much the same way that two numbers can be combined to get another number (for instance, 2 + 3 = 5), two isometries can be combined to get another isometry. High school students explore these combinations, discovering, for example, that if an object is successively reflected across two parallel lines, the first and final objects are related by a single translation. Likewise, they find that if an object is successively reflected across two intersecting lines, the first and final objects are related by a single rotation. These situations are illustrated in figure 3.7. In part (a) of the figure, object A is successively reflected across parallel lines r and s, to get object B. Note that, for example, point X on object A is reflected across line r to get point X', which is then reflected across line s to get point $X'' = Y$ on object B. Object A, including point X on object A, can be directly transformed to object B, including point X'', by the translation defined by the vector PQ. In part (b), object C is successively reflected across intersecting lines m and n to get object D. Object C can be directly transformed to get object D by a rotation with center P.

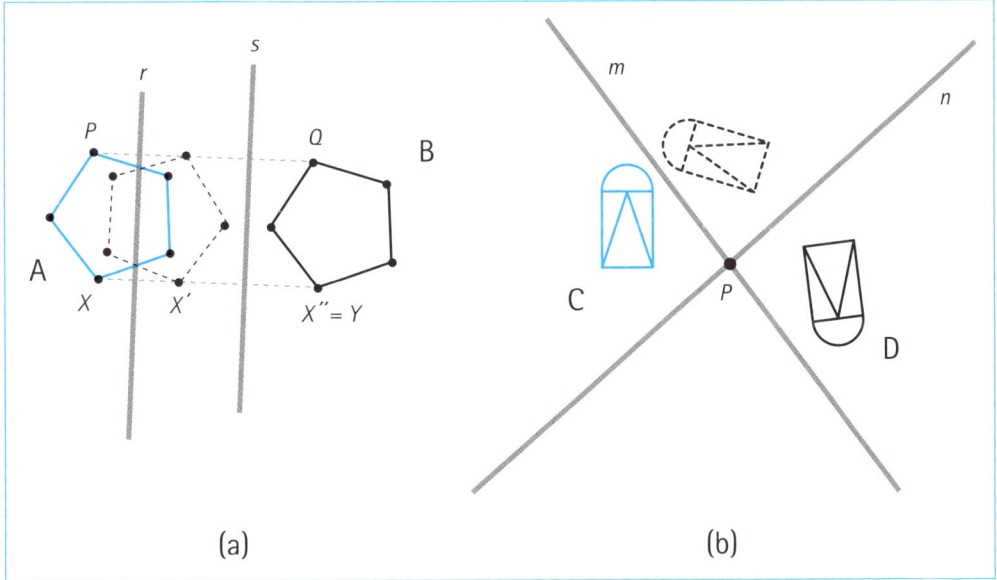

Fig. 3.7. Reflecting an object successively over (a) parallel lines and (b) intersecting lines. In both cases, the final object is related to the original object by a single isometry—a translation in (a) and a rotation in (b).

In grades 9–12, students' investigations show them that considering geometry from a transformational viewpoint sometimes makes seeing why certain theorems are true very easy. For example, in figure 3.8, parallelogram *ABDC* is composed of △*ABC* and the triangle formed by rotating it 180 degrees around the midpoint *M* of one of its sides. Because △*ABC* and △*DCB* are congruent, students can immediately see why the opposite angles of a parallelogram are congruent and why its opposite sides are congruent as well.

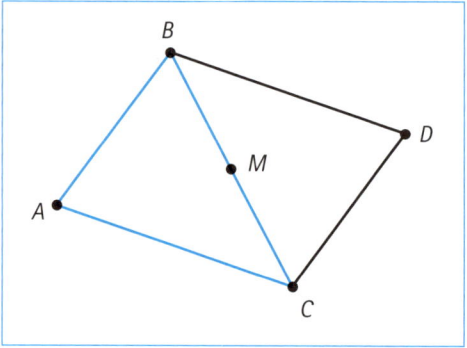

Fig. 3.8. A parallelogram composed of △*ABC* and the triangle formed by rotating △*ABC* 180 degrees

Explorations of transformations in grades 9–12 also provide students with insights into connections between geometry and algebra. Because isometries and dilations are nothing more than functions, transformations demonstrate that geometry and algebra do not need to be artificially separated but can be considered as parts of an integrated whole. As illustrated in figure 3.9, for example, in the Cartesian coordinate plane, a reflection across the *x*-axis maps a point with coordinates (*a*, *b*) to a point with coordinates (*a*, –*b*). Similarly, a reflection across the *y*-axis maps a point with coordinates (*a*, *b*) to a point with coordinates (–*a*, *b*). Algebraically, a reflection across the *x*-axis can be represented by the function *f*: (*a*, *b*) → (*a*, –*b*) and a reflection across the *y*-axis can be represented by the function *g*: (*a*, *b*) → (–*a*, *b*).

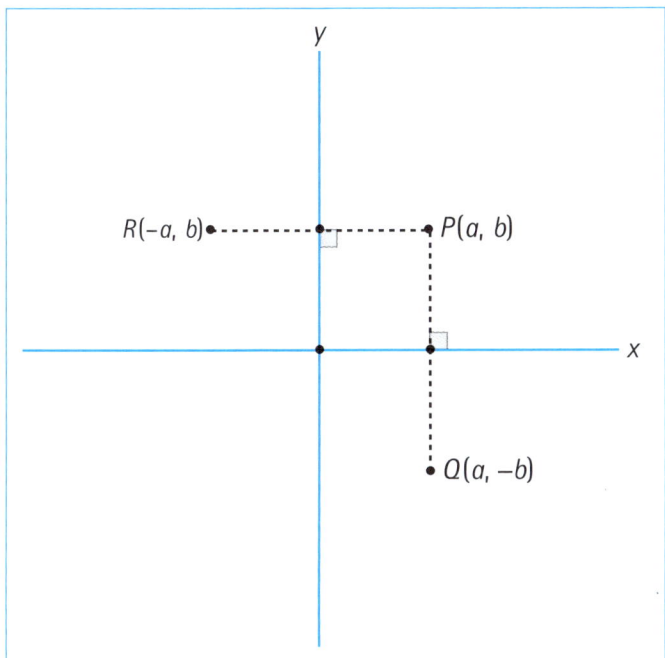

Fig. 3.9. The point *P* with Cartesian coordinates (*a*, *b*) and its reflections *R* and *Q* across the *y*- and *x*-axes, respectively, along with their coordinates

High school students who are given the translation shown in figure 3.10, defined by the vector *RS*, recognize that the point *P* with coordinates (*a*, *b*) is mapped to the point *Q* with coordinates (*a* + *m*, *b* + *n*), with $\frac{n}{m}$ as the slope of $\overline{RS}$. They understand, therefore, that this translation can be represented by the function *h*: (*a*, *b*) → (*a* + *m*, *b* + *n*) and that the translation is completely described by the vector: The translation that maps (*a*, *b*) to (*a* + *m*, *b* + *n*) is defined by the vector whose slope is $\frac{n}{m}$ and whose magnitude can be calculated by using the Pythagorean theorem to be $\sqrt{n^2 + m^2}$.

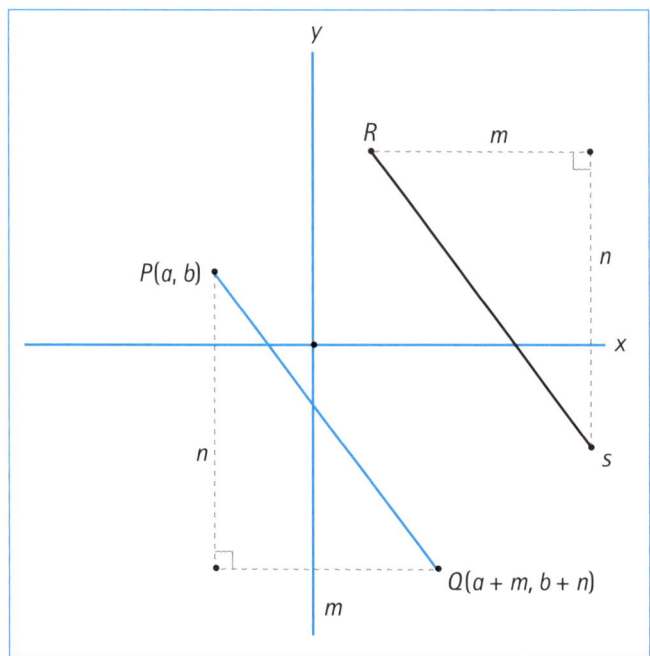

Fig. 3.10. The translation that maps (a, b) to ($a + m$, $b + n$), defined by the vector with slope $\frac{n}{m}$ and magnitude $\sqrt{n^2 + m^2}$

In grades 9–12, students also discover that other transformations exist that do not qualify as either an isometry or a dilation. These transformations preserve neither size nor shape. Students learn that one such transformation is a *shear*—a transformation of the plane in which all the points along a given line *m* remain fixed while the other points are shifted parallel to *m* by a distance proportional to their perpendicular distance from *m*. In figure 3.11, for example, square A is sheared parallel to the *x*-axis, resulting in parallelogram B. All the points on the *x*-axis are fixed, whereas all the points on lines parallel to the *x*-axis are shifted the same distance to the right (or left).

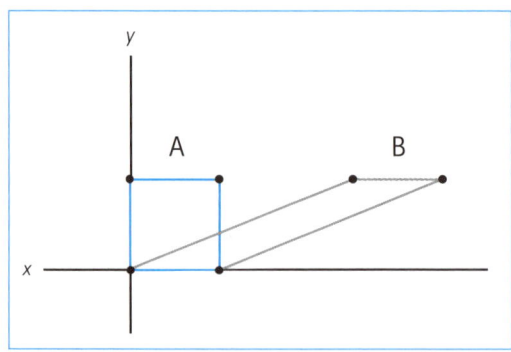

Fig. 3.11. A shear of square A parallel to the x-axis, resulting in parallelogram B

Figure 3.12 shows a square of side *r*, sheered by scaling factor *k*. Students discover that the shear with scaling factor *k* maps the point with coordinates (*a*, *b*) to the point with coordinates (*a* + *kb*, *b*). Applying insights from algebra, students understand that this shear can thus be represented by the function *l*: (*a*, *b*) → (*a* + *kb*, *b*). Students also learn that, although shears preserve neither size nor shape, they do preserve area. The square and parallelogram in figure 3.12 have the same areas, since they have the same base and the same height.

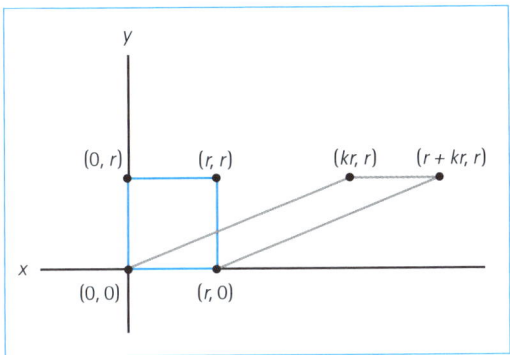

Fig. 3.12. A square of side *r*, sheared by scaling factor *k*

Students learn that the fundamental ideas that they explored when they were finding the perimeter and area of polygons such as triangles, parallelograms, rectangles, and squares can be useful when they are trying to find the perimeter and area of more complicated polygons or curvilinear figures such as circles and ellipses. They can develop the formula for the circumference of a circle experimentally by wrapping a string around various circular objects and then comparing the length of the string to the diameters of the objects. Within measurement error, this comparison is a constant approximately equal to 6.3, which leads to the formula $C = \pi d = 2\pi r$. In addition, as students discover, knowing how to find the area of an isosceles triangle proves to be very useful when they are seeking the area of any regular polygon—say, for example, the regular decagon shown in figure 3.13.

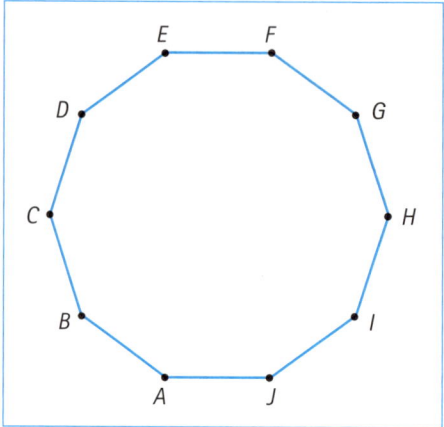

Fig. 3.13. A regular decagon

Students in grades 9–12 can reason about the area of the regular decagon as follows: A regular n-sided polygon can be divided into n congruent isosceles triangles, as illustrated for the decagon by figure 3.14. Therefore, finding the area of the regular polygon is equivalent to finding the area of an isosceles triangle and then multiplying by n. For the regular decagon in the figure, the area of $\triangle OAJ$ is

$$\frac{1}{2}as,$$

where a is the height of $\triangle OAJ$ (called the *apothem*), and s is the length of one of the sides of the decagon. So, for the regular decagon, the area is given by

$$10 \times \left(\frac{1}{2}as\right) = 5as.$$

Thus, students can reason that, for an n-sided regular polygon, the formula becomes

$$A = \frac{1}{2}ap$$

where a is the value of the apothem and p is the length of the polygon's perimeter.

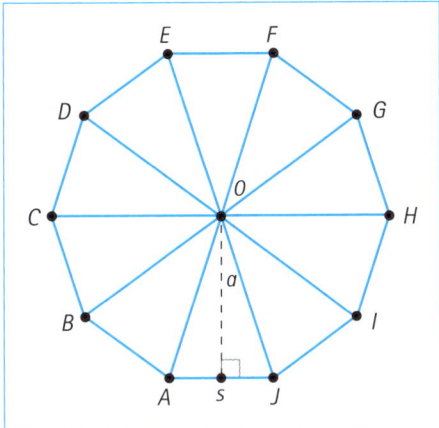

Fig. 3.14. A regular decagon divided into ten isosceles triangles

High school students can probe further, reasoning that as the number of sides increases, a regular polygon more closely resembles a circle, as illustrated in figure 3.15. They can then generalize the formula for the area of a circle from the formula for the area of a regular polygon. As the value of n grows infinitely large, the apothem of the regular polygon becomes the radius of the circle, and the perimeter of the regular polygon becomes the circumference of the circle. Hence, the formula

$$A = \frac{1}{2}ap \text{ becomes } A = \frac{1}{2}r(2\pi r) = \pi r^2.$$

CCSSM expects high school students to be able to reason in this way (NGA Center and CCSSO 2010, p. 78):

> Give an informal argument for the formulas for the circumference of a circle, area of a circle, volume of a cylinder, pyramid, and cone. *Use dissection arguments, Cavalieri's principle, and informal limit arguments.* (G-GMD.1.)

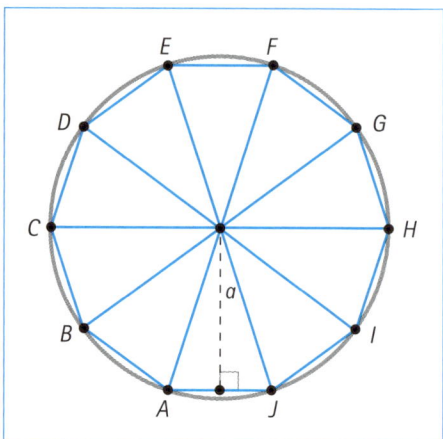

Fig. 3.15. As the number of sides increases in a regular polygon, its perimeter more nearly approaches the circumference of a circle.

Dynamic geometry software packages, such as the Geometer's Sketchpad or GeoGebra, are powerful tools that can help high school students understand complicated geometric properties. This software can measure many quantities, including length, angles, perimeter, area, and slope, and these capabilities allow students to explore the results of changes made to geometric figures, transformations, and constructions. Students who otherwise might struggle with geometry sometimes engage more easily in geometric discovery when they have the benefit of the dynamic nature of the software.

As an example of the power of dynamic geometry software (DGS), consider the situation in figure 3.16. A DGS application has been used to create a dilation centered at point P with scaling factor 0.5 that maps $\triangle BCD$ to $\triangle B'C'D'$. The DGS application has also calculated the areas of the two triangles, as well as the ratio of the areas.

To investigate the relationship between the areas, students can move $\triangle BCD$ around while changing its shape and size. If students hold the scaling factor at 0.5 so that the ratio of the areas remains constant, a "nice" relationship is likely to become apparent to them. Other examples, including the one shown in figure 3.17, show students that the ratio of the areas remains 0.25 for all triangles under a dilation with scaling factor 0.5. Investigating the consequences of using other scaling factors can lead students to discover that the ratio of the triangles' areas is the square of the scaling factor. Regardless of whether students discover this relationship, the dynamic environment gives all students opportunities to explore and make conjectures that they would not otherwise have been able to pursue.

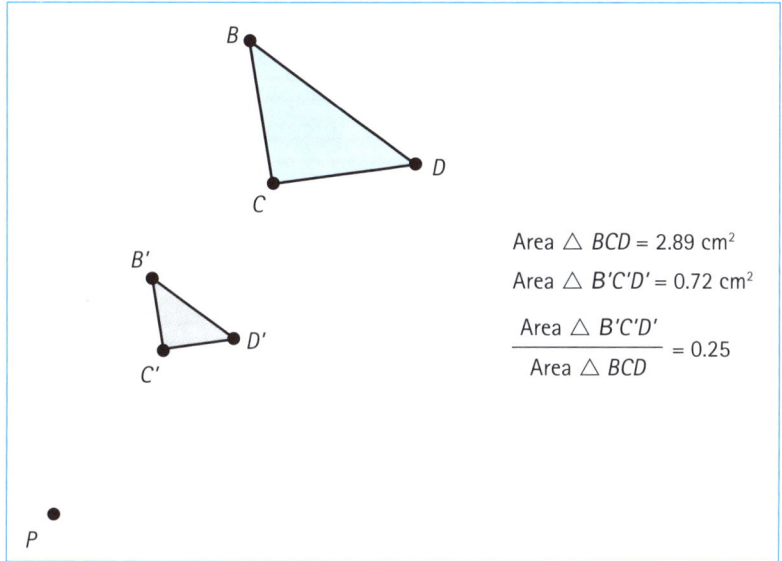

Area △ BCD = 2.89 cm²

Area △ B'C'D' = 0.72 cm²

$$\frac{\text{Area } \triangle\ B'C'D'}{\text{Area } \triangle\ BCD} = 0.25$$

Fig. 3.16. A dilation performed by a DGS application, with accompanying calculations

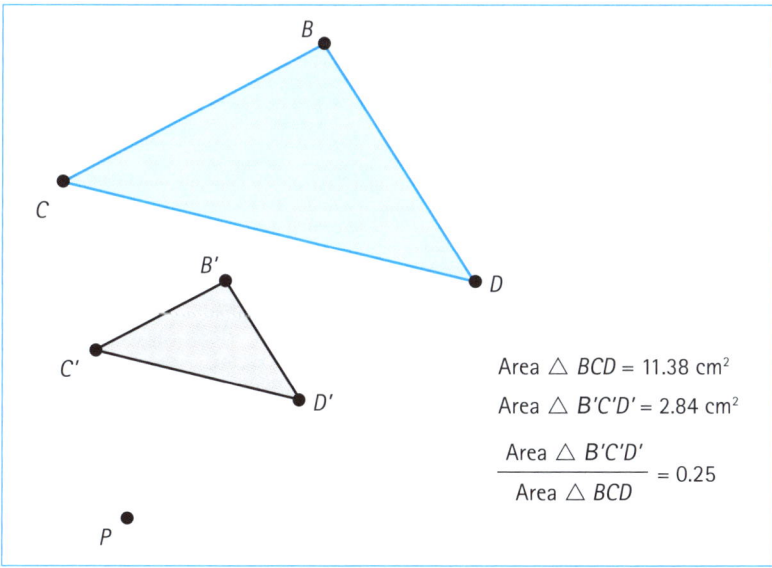

Area △ BCD = 11.38 cm²

Area △ B'C'D' = 2.84 cm²

$$\frac{\text{Area } \triangle\ B'C'D'}{\text{Area } \triangle\ BCD} = 0.25$$

Fig. 3.17. Exploring another triangle BCD under a dilation with a scaling factor of 0.5

Conclusion

Students come to grades 6–8 with strong foundations in geometry from their elementary school years. In grades 3–5, they have reasoned informally with shapes and their attributes, explored area, perimeter, and volume of rectangular prisms, and developed spatial sense that allows them to compose and decompose 2- and 3-dimensional shapes. Geometric measurement and transformational geometry in the middle grades build on this work and move students forward, strengthening crucial foundational concepts and skills. In grades 9–12, as students gain the understanding that reflections, rotations, and translations can be used to show congruence, they can extend their understanding of those relationships to consider similarity by performing dilations. Building knowledge in this way across the grades empowers students to use geometry in algebraic contexts, and vice versa, preparing students for work with geometry in college, careers in a variety of fields, and everyday life.

Appendix 1
The Big Ideas and Essential Understandings for Geometry

This book focuses on big ideas and essential understandings that are identified and discussed in *Developing Essential Understanding of Geometry for Teaching Mathematics in Grades 6–8* (Sinclair, Pimm, and Skelin 2012). For the reader's convenience, the full list of the big ideas and essential understandings in that book is reproduced below. The big ideas and essential understandings that are the special focus of this book are highlighted in blue.

Big Idea 1. Behind every measurement formula lies a geometric result.

Essential Understanding 1a. Decomposing and rearranging provide a geometric way of both *seeing that* a measurement formula is the right one and *seeing why* it is the right one.

Essential Understanding 1b. In addition to decomposing and rearranging, shearing provides another geometric way of both *seeing that* a measurement formula is the right one and *seeing why* it is the right one.

Big Idea 2. Geometric thinking involves developing, attending to, and learning how to work with imagery.

Essential Understanding 2a. Geometric images provide the content in relation to which properties can be noticed, definitions can be made, and invariances can be discerned.

Essential Understanding 2b. Symmetry provides a powerful way of working geometrically.

Essential Understanding 2c. Geometric awareness develops through practice in visualizing, diagramming, and constructing.

Big Idea 3. A geometric object is a mental object that, when constructed, carries with it traces of the tool or tools by which it was constructed.

Essential Understanding 3*a*. Tools provide new sources of imagery as well as specific ways of thinking about geometric objects and processes.

Essential Understanding 3*b*. Geometric thinking turns tools into objects, and in geometry the process of turning an action undertaken with a tool into an object happens over and over again.

Big Idea 4. Classifying, naming, defining, posing, conjecturing, and justifying are codependent activities in geometric investigation.

Essential Understanding 4*a*. Naming is not just about nomenclature: it draws attention to properties and objects of geometric interest.

Essential Understanding 4*b*. Definition can both generate and reflect structure: definitions are often dependent on a specific classification.

Essential Understanding 4*c*. Conjectures can emerge out of a problem-posing process that generates claims that need to be justified.

Appendix 2
Resources for Teachers

The following list highlights a few of the many books, articles, and websites that are helpful resources for teaching geometry in grades 6–8. Abstracts from the publishers provide brief descriptions of the resources.

Books

Battista, Michael T. *Cognition-Based Assessment and Teaching of Geometric Measurement: Building on Students' Reasoning.* Portsmouth, N.H.: Heinemann, 2012.

——. *Cognition-Based Assessment and Teaching of Geometric Shapes: Building on Students' Reasoning.* Portsmouth, N.H.: Heinemann, 2012.

Using a research-based framework that describes the development of students' thinking and learning in terms of levels of sophistication, a "cognitive terrain" that includes ascents and plateaus, Battista shows how teachers can build on their students' reasoning. His approach emphasizes three key components that support students' mathematical sense making and proficiency:

- Determining students' levels of sophistication in reasoning

- Assessing and monitoring the development of students' understanding of core ideas

- Differentiating instruction to meet individual students' learning needs

Driscoll, Mark. *Fostering Geometric Thinking: A Guide for Teachers, Grades 5–10.* Portsmouth, N.H.: Heinemann, 2007.

Anyone teaching geometry can discover essential, practical ideas in this book for helping students cultivate geometric habits of mind that lead to success with this core mathematical topic. The author focuses on rigorous, problem-based teaching that encourages students to deepen their thinking in three key geometric strands:

- Geometric properties

- Geometric transformations

- Measurement of geometric objects

Discover how the interplay of these strands supports students in devising multiple solutions and developing a broader sense of geometric principles.

Sinclair, Nathalie, David Pimm, and Melanie Skelin. *Developing Essential Understanding of Geometry for Teaching Mathematics in Grades 6–8*. Essential Understanding Series. Reston, Va.: National Council of Teachers of Mathematics, 2012.

> This book focuses on essential knowledge for mathematics teachers about geometry and measurement. It is organized around four big ideas, supported by multiple smaller, interconnected ideas—*essential understandings*. Taking teachers beyond a simple introduction to geometry, the authors aim to broaden and deepen teachers' understanding of one of the most challenging mathematical topics for students—and teachers. Developing this understanding will help teachers engage students, anticipate their perplexities, avoid pitfalls, and dispel misconceptions. The book also suggests how to develop appropriate tasks, techniques, and tools for assessing students' understanding of the topics.

Usiskin, Zalman, and Jennifer Griffin. *The Classification of Quadrilaterals: A Study of Definition*. Charlotte, N.C.: Information Age, 2008.

> This book reports on an analysis of a small part of the mathematics curriculum—the definitions given to quadrilaterals. Some disagreement exists about the definitions and, consequently, about the ways in which quadrilaterals are classified and relate to one another. The issues underlying these differences perennially engage students, teachers, mathematics educators, and mathematicians. Numerous articles and essays have addressed the definitions and classification of quadrilaterals, and curricular materials reflect the broad mathematical issues revolving around definitions. The intended audience for this book includes curriculum developers, researchers, teachers, teacher trainers, and anyone interested in language and its use.

Van de Walle, John, Karen S. Karp, and Jennifer M. Bay-Williams. *Elementary and Middle School Mathematics: Teaching Developmentally*. 7th ed. Allyn & Bacon, 2010.

> The authors wrote this book to help teachers understand mathematics and become confident in their ability to teach the subject to children in kindergarten through eighth grade. The chapters related to the teaching and learning of geometric thinking, geometric concepts, and measurement concepts provide ideas and insights that can support teachers as they design and implement their lessons.

Articles

Bell, Carol J. "Measuring Tangrams on a Geoboard." *Mathematics Teaching in the Middle School* 22 (February 2017): 374–78.

> The author provides a geometric task that involves using all seven of the tangram pieces to produce a shape on the geoboard and then finding its perimeter and area. This activity offers opportunities for developing measuring skills as well as identifying patterns in tables.

Canada, Daniel L., Matthew A. Ciancetta, and Stephen D. Blair. "Going Off-the-Pegs: Geoboard Squares." *Mathematics Teaching in the Middle School* 20 (December 2014): 286–92.

> In a lively investigation, students consider what happens when vertices of squares are not on the pegs of a geoboard.

Cady, Jo Ann, and Pamela Wells. "Odd Shape Out." *Mathematics Teaching in the Middle School* 22 (September 2016): 72–76.

> The authors present a task in which students are shown four quadrilaterals and are asked to find the shape that does not fit in the group. As an extension of their work on this task, students are asked to find a way in which any of the four quadrilaterals could be the "odd shape out."

Dimmel, Justin K., and Patricio G. Herbst. "The Semiotic Structure of Geometry Diagrams: How Textbook Diagrams Convey Meaning." *Journal for Research in Mathematics Education* 46 (March 2015): 147–95.

> Noting that geometry diagrams use the visual features of specific drawn objects to convey meaning about generic mathematical entities, the authors examine the semiotic structure of these visual features in two parts. First they conduct a semiotic inquiry to conceptualize geometry diagrams as mathematical texts that incorporate choices from different semiotic systems. Then they use the semiotic catalog that results from this inquiry to analyze 2,300 diagrams from twenty-two high school geometry textbooks with publication dates spanning the twentieth century. In the first part of the article, the authors identify axes along which the features of geometry diagrams can vary, and in the second part they show the viability of using the semiotic framework to conduct empirical studies of diagrams in geometry textbooks.

Garimella, Umadevi I., and Belinda M. Robertson. "Modeling the Shapes of Cells." *Mathematics Teaching in the Middle School* 21 (October 2015): 180–88.

> The authors show how students can extend a previous task in which they worked with tiles to explore perimeter and area relationships to engage in a new task in which they investigate relationships between surface area and volume.

Haberern, Colleen. "The Cake Contest." *Mathematics Teaching in the Middle School* 22 (December 2016): 274–82.

> The author shares a way to incorporate a 3-D printer with tasks involving surface area and volume. The article includes samples of student solution methods along with the task.

Jeon, Kyungsoon. "Mathematics Hiding in the Nets for a Cube." *Teaching Children Mathematics* 15 (March 2009): 394–99.

> The author expands on a mathematical discussion among third graders in response to the question, "How many different nets can you draw that can fold into a cube?" The article also makes connections to other platonic solids and Euler's formula.

Keiser, Jane M., Amanda Klee, and Karen Fitch. "An Assessment of Students' Understanding of Angle." *Mathematics Teaching in the Middle School* 9 (October 2003): 116–19.

> The authors discuss assessment tasks related to angles and classify students' definitions of angles.

Lehrer, Richard, Marta Kobiela, and Paul J. Weinberg. "Cultivating Inquiry about Space in a Middle School Mathematics Classroom." *ZDM* 45, no. 3 (2013): 365–76.

In forty-six lessons in Euclidean geometry, sixth-grade students (ages 11 and 12) were initiated into the use of the mathematical practice of inquiry. Teachers supported inquiry by soliciting student questions and orienting students to the use of such interrelated mathematical habits of mind as generalizing, developing relations, and seeking invariants in light of change, to sustain investigations of their questions. In a comparison of earlier and later phases of instruction, the researchers found that the questions posed by students reflected their increasing disposition to seek generalization and explore mathematical relationships—ways of thinking valued by the mathematics discipline. Less prevalent were questions directed toward seeking invariants in light of change. But when students did pose such questions, they tended to be oriented toward generalizing and establishing relations among mathematical objects and properties. As instruction proceeded, students developed an aesthetic that emphasized the value of questions oriented toward the collective pursuit of knowledge. Post-instructional interviews with the students revealed that they experienced the forms of inquiry and investigation cultivated in the classroom as self-expressive.

Milliman, Wanda. "It's a Bird, It's a Plane, It's a Dilated Superhero." *Mathematics Teaching in the Middle School* 21 (April 2016): 498–504.

The author describes students' engagement in a dilation task that also involves them in graphing and describing a rule algebraically. The article includes activity sheets for the task.

Prummer, Kathy E., Julie M. Amador, and Abraham J. Wallin. "Persevering with Prisms: Producing Nets." *Mathematics Teaching in the Middle School* 21 (April 2016): 472–79.

The authors use three essential understandings from *Developing Essential Understanding of Geometry for Teaching Mathematics in Grades 6–8* (Sinclair, Pimm, and Skelin 2012) as the foundation for this article. They focus on using nets of rectangular prisms, then scaling them to produce nets that result in prisms with a given volume.

Roscoe, Matt B., and Joe Zephyrs. "Quilt Block Symmetries." *Mathematics Teaching in the Middle School* 22 (August 2016): 18–27.

The authors show how transformational geometry can provide a way to explore congruence, symmetry, and similarity in the context of quilts. They describe an investigation in which students have opportunities to create justifications for their solutions and critique those of others.

Teuscher, Dawn, Dung Tran, and Barbara J. Reys. "Common Core State Standards in the Middle Grades: What's New in the Geometry Domain and How Can Teachers Support Student Learning?" *School Science and Mathematics* 115, no. 1 (2015): 4–13.

The Common Core State Standards for Mathematics (CCSSM) are a primary focus of attention for many stakeholders who are intent on improving mathematics education, including teachers, district mathematics leaders, and curriculum developers. This article reports on specific content shifts in the geometry domain in the middle grades (6–8) mathematics curriculum. The authors employ a methodology that allows them to compare content across multiple standards documents, and they identify dramatic changes with respect to the geometry content taught in the middle grades. In an analysis of eight states, they found that 52% of CCSSM's middle grades geometry standards expect students to develop skills and concepts at a particular grade level that they were not previously expected to master at that level (57% in grade 6, 50% in grade 7, and 50% in grade 8). On the basis of their analysis of CCSSM and pre-CCSSM state standards, the authors highlight three areas that represent "new" geometry content at the middle grades.

Online Resources

NCTM Illuminations Lessons

http://illuminations.nctm.org/

The NCTM Illuminations project originated as part of the Verizon Thinkfinity program and is currently maintained by NCTM. The website presents a variety of standards-based resources, including lessons, activities, and hundreds of Web links.

Illustrative Mathematics Project

http://www.illustrativemathematics.org

Illustrative Mathematics provides guidance to states, assessment consortia, testing companies, and curriculum developers by illustrating the range and types of mathematical work that support implementation of the Common Core State Standards. One tool on this website is a growing collection of mathematical tasks that are organized by standard for each grade level and illustrate important features of the indicated standard or standards. The tasks on the website are not meant to be considered in isolation. Taken together in sets, these tasks are intended to illustrate a particular standard. Eventually, the site will showcase sets of tasks for each standard that—

- illuminate the central meaning of the standard and show connections with other standards;

- clarify what is familiar about the standard and what is new with the advent of the Common Core State Standards;

- include both teaching and assessment tasks; and

- reflect the full range of difficulty that the standard expects students to master.

Appendix 3
Tasks

This book examines rich tasks that have been used in the classroom to bring to the surface students' understandings and misunderstandings about geometry. A sampling of these tasks is presented here, in the order in which they appear in the book:

Creating Rectangles with a Given Area or Perimeter

Working with a Nonuniform Shape: The Leaf Task

Finding the Nets of a Cube

Conjecturing about the Relationship between the Volumes of Two Cubes

Investigating Nets: Open Box vs. Cube

Reflecting Three Given Objects

Focusing on the Points in a Rotation

Identifying the Center of a Rotation

Reasoning about Similar Triangles

Making Sense of Ratios in a Dilation

At More4U, Appendix 3 includes all the tasks discussed in the book, formatted for classroom use and ready for printing.

Creating Rectangles with a Given Area or Perimeter

Task A

Use square tiles to create as many different rectangles as possible, each with an area of 36 square units. Draw your rectangles on grid paper.

Task B

Use square tiles to create as many different rectangles as possible, each with a perimeter of 24 units. Draw your rectangles on grid paper.

Working with a Nonuniform Shape: The Leaf Task

(Adapted by Wilson and Chavarria [1993] from Ronau and Gilbert [1988])

Russell's group and Jordan's group were asked to find the area of the leaf shown below.

Decide whether each group used accurate methods, and explain your thinking.

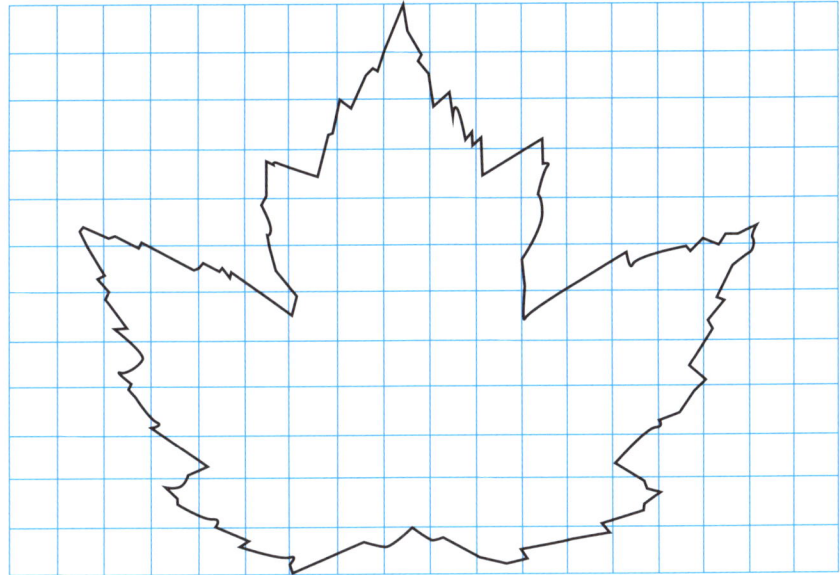

- Russell's group found the area of the leaf by counting all the squares inside the leaf's boundary. His group paired half squares to create full square units and totaled all the square units.
- Jordan's group took a string and placed it closely around the perimeter of the leaf. Then they created a rectangle with the measured string and counted the length and width of the shape in sides of squares and multiplied them to find the area.

Finding the Nets of a Cube

Below is a net for a cube:

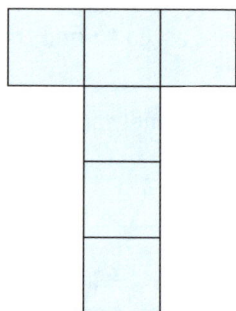

How many distinct nets can be made for a cube?

Conjecturing about the Relationship between the Volumes of Two Cubes

The edge of a cube measures 5 cm. Another cube has an edge length of 10 cm. Without performing any calculations, make a conjecture about the relationship between the volumes of the two cubes.

Also make a conjecture about the relationship between the surface areas of the two cubes. Justify your answers.

Investigating Nets: Open Box vs. Cube

Below is a two-dimensional representation of an open-top box:

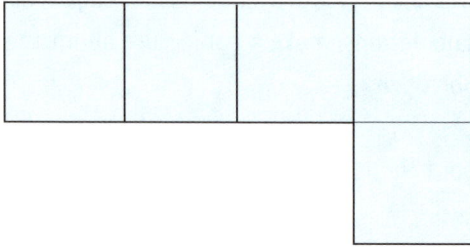

a. How many ways can you add a square to create a net for a cube?

b. Find three other two-dimensional representations that will fold into an open-top box.

c. How many ways can you add a square to your other representations to create a net for a cube?

d. What observations can you make?

Reflecting Three Given Objects

Reflect objects Q, R and H over line *m*. What do you notice?

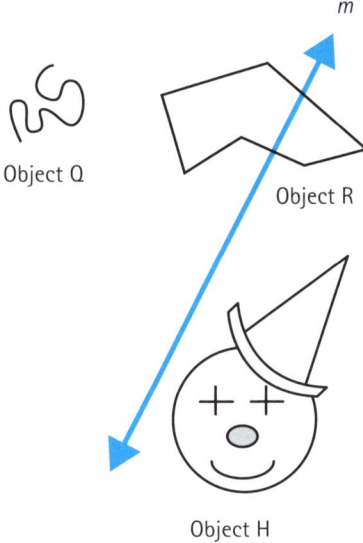

Reflect objects Q, R and H over line *m*.

Focusing on the Points in a Rotation

A rotation of 120° around point *P* is illustrated below. Do any points remain in their original positions?

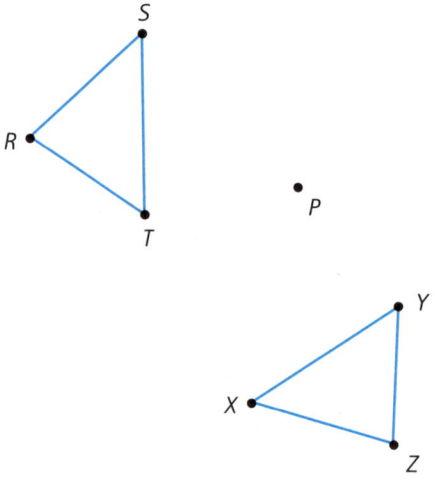

Identifying the Center of a Rotation

Hexagon *A′B′C′D′E′F′* is the image of hexagon *ABCDEF* under a rotation of the plane of 200°. Which of the points *V, W, X, Y,* or *Z* is the center of this rotation?
Explain your reasoning.

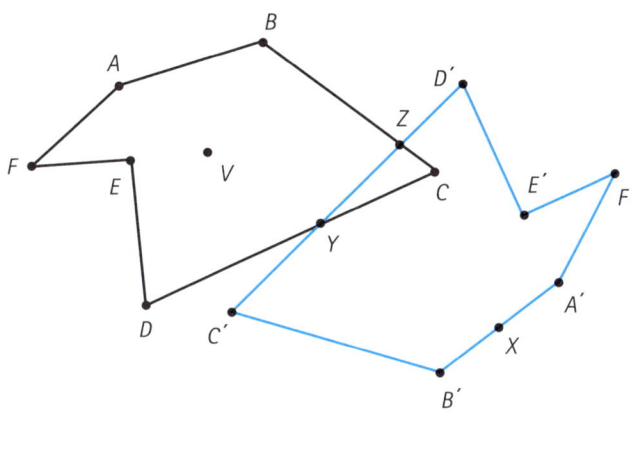

Reasoning about Similar Triangles

What is the length of $\overline{AC}$ in similar triangles ABC and $A'B'C'$? Monte says it is

$13\frac{1}{3}$ inches, but Keena claims it is 12. Keena explains her reasoning:

"Since 10 = 6 + 4, the length of $\overline{AC}$ must be 8 + 4, or 12 inches."

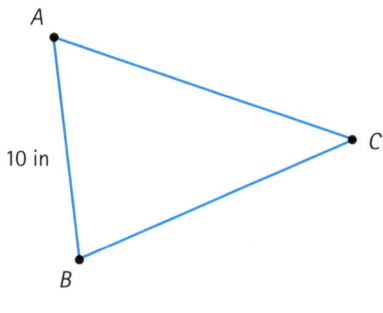

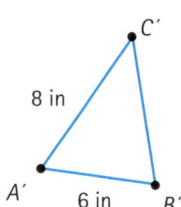

Do you agree with Keena's explanation?

What would you say to Keena and Monte?

Making Sense of Ratios in a Dilation

Two students, Hannah and Gina, are discussing the similar pentagons *ABCDE* and *A'B'C'D'E'*, shown below.

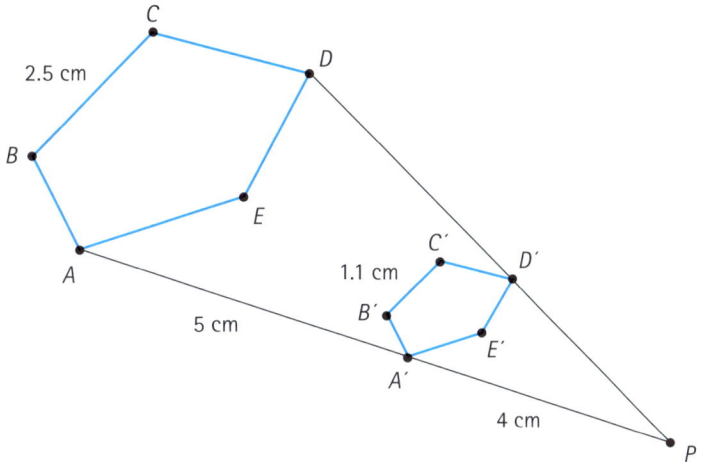

Gina is perplexed and says that something must be wrong with the labeling of the lengths.

"We know the pentagons are similar," she says. "So corresponding measurements must be proportional. But $\frac{4}{5}$ is not even close to $\frac{1.1}{2.5}$, even if some of the measurements are rounded!"

Hannah disagrees. "The measurements look OK to me," she says.

Who is correct, Gina or Hannah? Why?

How would you explain this to them?

References

Barrett, Jeffrey E., Craig Cullen, Julie Sarama, Douglas H. Clements, David Klanderman, Amanda Miller, and Chepina Rumsey. "Children's Unit Concepts in Measurement: A Teaching Experiment Spanning Grades 2 through 5." *ZDM Mathematics Education* 43 (September 2011): 637. doi:10.1007/s11858-100-0368-8.

Battista, Michael T. "Learning Geometry in a Dynamic Computer Environment." *Teaching Children Mathematics* 8 (February 2002): 333–39.

———. "Understanding Students' Thinking about Area and Volume Measurement." In *Learning and Teaching Measurement*, Sixty-Fifth Yearbook of the National Council of Teachers of Mathematics (NCTM), edited by Douglas H. Clements, pp. 122–42. Reston, Va.: NCTM, 2003.

———. "Understanding the Development of Students' Thinking about Length." *Teaching Children Mathematics* 13 (October 2006): 140–46.

Burrill, John, ed. *Mathematics Assessment Sampler, Grades 6-8*. Reston, Va.: National Council of Teachers of Mathematics, 2005.

Clements, Douglas H., and Julie Sarama. "Early Childhood Teacher Education: The Case of Geometry." *Journal of Mathematics Teacher Education* 14 (February 2011): 133–48.

de Hevia, Maria D., Giuseppe Vallar, and Luisa Girelli. "Visualizing Numbers in the Mind's Eye: The Role of Visuo-Spatial Processing in Numerical Abilities." *Neuroscience and Biobehavioral Reviews* 32 (October 2008): 1361–72.

Dougherty, Barbara J. "Access to Algebra: A Process Approach." In *The Future of the Teaching and Learning of Algebra*, edited by Helen Chick, Kay Stacey, Jill Vincent, and John Vincent, pp. 207–13. Victoria, Australia: University of Melbourne, 2001.

———. *The Write Way: Journal Prompts and More, Grades 5 and 6*. Honolulu, Hawaii: University of Hawaii, Curriculum Research and Development Group, 2006.

Edwards, Laurie. D. "Children's Learning in a Computer Microworld for Transformational Geometry." *Journal for Research in Mathematics Education* 22 (March 1991): 122–37.

Garimella, Umadevi I., and Belinda M. Robertson. "Modeling the Shapes of Cells." *Mathematics Teaching in the Middle School* 21 (October 2015): 180–88.

Grossman, Pamela. *The Making of a Teacher*. New York: Teachers College Press, 1990.

Haberern, Colleen. "The Cake Contest." *Mathematics Teaching in the Middle School* 22 (December 2016): 274–82.

Hattie, John. *Visible Learning: A Synthesis of Over 800 Meta-Analyses Relating to Achievement.* New York: Routledge, 2009.

Hiebert, James. "Units of Measure: Results and Implications from National Assessment." *Arithmetic Teacher* 28 (February 1981): 38–43.

Hill, Heather C., Brian Rowan, and Deborah Loewenberg Ball. "Effects of Teachers' Mathematical Knowledge for Teaching on Student Achievement." *American Educational Research Journal* 42 (Summer 2005): 371–406.

Jeon, Kyungsoon. "Mathematics Hiding in the Nets for a Cube." *Teaching Children Mathematics* 15 (March 2009): 394–99.

Jones, Keith. "Critical Issues in the Design of the Geometry Curriculum." In *Readings in Mathematics Education*, edited by Bill Barton, pp. 75–90. Auckland, New Zealand: University of Auckland, 2000.

Kamii, Constance. "Measurement of Length: How Can We Teach It Better?" *Teaching Children Mathematics* 13 (October 2006): 154–58.

Kline, Morris. *Mathematical Thought from Ancient to Modern Times*. Vol. 1. New York: Oxford University Press, 1972.

Küchemann, Dietmar. "Reflections and Rotations." In *Children's Understanding of Mathematics: 11–16*, edited by Kathleen M. Hart, pp. 137–57. London: John Murray, 1981.

Lamberg, Teruni. *Whole Class Mathematics Discussions: Improving In-Depth Mathematical Thinking and Learning*. Boston: Allyn & Bacon, 2012.

Lindquist, Mary M., and Douglas H. Clements. "Geometry Must Be Vital." *Teaching Children Mathematics* 7 (March 2001): 408–15.

Magnusson, Shirley, Joseph Krajcik, and Hilda Borko. "Nature, Sources, and Development of Pedagogical Content Knowledge for Science Teaching." In *Examining Pedagogical Content Knowledge*, edited by Julie Gess-Newsome and Norman G. Lederman, pp. 95–132. Dordrecht, The Netherlands: Kluwer Academic, 1999.

Martin, W. Gary. "Supporting Secondary School Students' Construction of Geometric Knowledge." In *New Directions in Research in Geometry*, edited by Annette R. Baturo, pp. 74–79. Brisbane, Australia: Queensland University of Technology, 1996.

Masters, Jessica. "Diagnostic Geometry Assessment Project Technical Report: Item Characteristics." In *inTASC* Report (August 2010). Accessed April 17, 2017. http://www.bc.edu/research/intasc/researchprojects/dga/dga_publications.shtml.

Mulligan, Joanne, Michael Mitchelmore, and Anne Prescott. "Case Studies of Children's Development of Structure in Early Mathematics: A Two-Year Longitudinal Study." In *Proceedings of the 29th Conference of the International Group for the Psychology of Mathematics Education*, edited by Helen L. Chick and Jill L. Vincent, pp. 1–9. Melbourne, Australia: Program Committee, 2005.

National Council of Teachers of Mathematics (NCTM). *Principles and Standards for School Mathematics*. Reston, Va.: NCTM, 2000.

National Governors Association Center for Best Practices, Council of Chief State School Officers (NGA Center and CCSSO). *Common Core State Standards for Mathematics. Common Core State Standards (College- and Career-Readiness Standards and K–12 Standards in English Language Arts and Math)*. Washington, D.C.: NGA Center and CCSSO, 2010. http://www.corestandards.org.

Outhred, Lynne N., and Michael Mitchelmore. "Young Children's Intuitive Understanding of Rectangular Area Measurement." *Journal for Research in Mathematics Education* 31 (February 2000): 144–67.

Popham, W. James. "Defining and Enhancing Formative Assessment." Paper presented at the CCSSO State Collaborative on Assessment and Student Standards FAST meeting, Austin, Tex., October 10–13, 2006.

Prummer, Kathy E., Julie M. Amador, and Abraham J. Wallin. "Persevering with Prisms: Producing Nets." *Mathematics Teaching in the Middle School* 21 (April 2016): 472–79.

Ronau, Robert. N., and Robert K. Gilbert. "Investigating Relations in Nature." *School Science and Mathematics* 88 (November 1988): 573–80.

Rose, Cheryl M., Leslie Minton, and Carolyn B. Arline. *Uncovering Student Thinking in Mathematics: 25 Formative Assessment Probes.* Thousand Oaks, Ca.: Corwin Press, 2007.

Senk, Sharon L. "Van Hiele Levels and Achievement in Writing Geometry Proofs." *Journal for Research in Mathematics Education* 20 (May 1989): 309–21.

Shea, Daniel L., David Lubinski, and Camilla P. Benbow. "Importance of Assessing Spatial Ability in Intellectually Talented Young Adolescents: A 20-Year Longitudinal Study." *Journal of Educational Psychology* 93 (September 2001): 604–14.

Shulman, Lee S. "Those Who Understand: Knowledge Growth in Teaching." *Educational Researcher* 15, no. 2 (1986): 4–14.

–––. "Knowledge and Teaching." *Harvard Educational Review* 57, no. 1 (1987): 1–22.

Sinclair, Nathalie, David Pimm, and Melanie Skelin. *Developing Essential Understanding of Geometry for Teaching Mathematics in Grades 6–8.* Essential Understanding Series. Reston, Va.: National Council of Teachers of Mathematics, 2012.

Skemp, Richard R. "Relational Understanding and Instrumental Understanding." *Arithmetic Teacher* 26 (September 1978): 9–15.

Slovin, Hannah. "Moving to Proportional Thinking." *Mathematics Teaching in the Middle School* 6 (September 2000): 58–60.

Slovin, Hannah, Linda Venenciano, Melanie Ishihara, and Cynthia Beppu. *Reshaping Mathematics for Understanding: Motion Geometry.* Honolulu: Curriculum Research and Development Group, University of Hawaii, 2003.

Smith, Margaret Schwan, and Mary Kay Stein. *5 Practices for Orchestrating Productive Mathematics Discussions.* Reston, Va.: National Council of Teachers of Mathematics, 2011.

Sowder, Judith, Diane Wearne, W. Gary Martin, and Marilyn Strutchens. "What Do Eighth-Grade Students Know about Mathematics? Changes over a Decade." In *Results and Interpretations of the 1990 through 2000 Mathematics Assessment of the National Assessment of Educational Progress*, edited by Peter Kloosterman and Frank K. Lester Jr., pp. 145–74. Reston, Va.: National Council of Teachers of Mathematics, 2004.

Stavridou, Fontini, and Domna Kakana. "Graphic Abilities in Relation to Mathematical and Scientific Ability in Adolescents." *Educational Research* 50 (March 2008): 75–93.

Steele, Michael. *Middle Grades Geometry and Measurement: Examining Change in Knowledge Needed for Teaching through a Practice-Based Teacher Education Experience.* PhD diss., University of Pittsburgh, 2006.

Stein, Mary Kay, Randi Engle, Margaret S. Smith, and Elizabeth K. Hughes. "Orchestrating Productive Mathematical Discussions: Helping Teachers Learn to Better Incorporate Student Thinking." *Mathematical Thinking and Learning* 10 (October 2008): 313–40.

Thompson, Alba G., Randolph A. Phillipp, Patrick W. Thompson, and Barbara A. Boyd. "Calculational and Conceptual Orientations in Teaching Mathematics." In *Professional Development for Teachers of Mathematics,* 1994 Yearbook of the National Council of Teachers of Mathematics (NCTM), edited by Douglas B. Aichele, pp. 79–92. Reston, Va.: NCTM, 1994.

Wai, Jonathan, David Lubinski, and Camilla P. Benbow. "Spatial Ability for STEM Domains: Aligning over 50 Years of Cumulative Psychological Knowledge Solidifies Its Importance." *Journal of Educational Psychology* 101 (November 2009): 817–35.

Wesslén, Maria, and Saínza Fernandez. "Transformation Geometry." *Mathematics Teaching* 191 (June 2005): 27–29.

Wiliam, Dylan. "Keeping Learning on Track: Classroom Assessment and the Regulation of Learning." In *Second Handbook of Research on Mathematics Teaching and Learning*, edited by Frank K. Lester Jr., pp. 1053–98. Charlotte, N.C.: Information Age; Reston, Va.: National Council of Teachers of Mathematics, 2007.

Wilson, Linda D., and Silvia Chavarria. "Superitem Test as a Classroom Assessment Tool." In *Assessment in the Mathematics Classroom*, 1993 Yearbook of the National Council of Teachers of Mathematics (NCTM), edited by Norman L. Webb, pp. 135–42. Reston, Va.: NCTM, 1993.

Yinger, Robert J. "The Conversation of Teaching: Patterns of Explanation in Mathematics Lessons." Paper presented at the meeting of the International Study Association on Teacher Thinking, Nottingham, England, May, 1988.

Young, John Wesley. *Lectures on Fundamental Concepts of Algebra and Geometry.* New York: Macmillan, 1911.

Titles in the Putting Essential Understanding into Practice Series

The Putting Essential Understanding into Practice Series takes NCTM's Essential Understanding Series to the next level through a focus on pedagogical content knowledge. Each volume builds on the companion volume in the earlier series to show teachers how to implement their understanding of mathematics in the classroom. The authors identify common misconceptions, along with strategies and activities to help students develop robust understanding through problem-based learning.

Putting Essential Understanding of—

Addition and Subtraction into Practice in Prekindergarten–Grade 2
ISBN 978-0-87353-730-8   Stock No. 14540

Geometry and Measurement into Practice in Prekindergarten–Grade 2
ISBN 978-0-87353-731-5   Stock No. 14541

Fractions into Practice in Grades 3–5
ISBN 978-0-87353-732-2   Stock No. 14542

Multiplication and Division into Practice in Grades 3–5
ISBN 978-0-87353-715-8   Stock No. 14347

Geometry and Measurement into Practice in Grades 3–5
ISBN 978-0-87353-733-9   Stock No. 14543

Ratios and Proportions into Practice in Grades 6–8
ISBN 978-0-87353-717-9   Stock No. 14349

Gcometry into Practice in Grades 6–8
ISBN 978-0-87353-734-6   Stock No. 14544

Functions into Practice in Grades 9–12
ISBN 978-0-87353-714-8   Stock No. 14346

Statistics into Practice in Grades 9–12
ISBN 978-0-87353-737-7   Stock No. 14547

Geometry into Practice in Grades 9–12
ISBN 978-0-87353-736-0   Stock No. 14546

Forthcoming:

Putting Essential Understanding of—

Number and Numerations into Practice in Prekindergarten–Grade 2

Expressions and Equations into Practice in Grades 6–8

Visit www.nctm.org/catalog for details and ordering information.

Titles in the Essential Understanding Series

The Essential Understanding Series gives teachers the deep understanding that they need to teach challenging topics in mathematics. Students encounter such topics across the pre-K–grade 12 curriculum, and teachers who understand the big ideas related to each topic can give maximum support as students develop their own understanding and make vital connections.

Developing Essential Understanding of–

Visit www.nctm.org/catalog for details and ordering information.